MW00607733

STATION
BLACKOUT

PHOTO BY TIFFANY CAPUANO

STATION BLACKOUT

INSIDE THE FUKUSHIMA NUCLEAR
DISASTER AND RECOVERY

CHARLES A. CASTO

RADIUS BOOK GROUP
NEW YORK

Distributed by Radius Book Group
A Division of Diversion Publishing Corp.
443 Park Avenue South, Suite 1004
New York, NY 10016
www.RadiusBookGroup.com

Copyright © 2018 by Charles A. Casto

All rights reserved, including the right to reproduce this book or portions
thereof in any form whatsoever.

Library of Congress Control Number: 2018954167

For more information, email info@radiusbookgroup.com.

First edition: December 2018
Hardcover ISBN: 978-1-63576-402-4
eBook ISBN: 978-1-63576-403-1

Manufactured in the United States of America
10 9 8 7 6 5 4 3 2

Jacket design by Mark Karis
Interior design by Pauline Neuwirth, Neuwirth & Associates

Radius Books and the Radius Books colophon
are registered trademarks of Radius Books, Inc.

To the heroes of Fukushima Daiichi and Daini

CONTENTS

STATION
BLACKOUT

INTRODUCTION

---

*"Taylor's not gone.... She's just away from us. Taylor's a
volunteer in heaven right now."*
—Sister of tsunami victim Taylor Anderson

f you are a leader, an engineer, a disaster buff, or someone personally connected to or touched by the horrific events that unfolded
on March 11, 2011—starting with the Great East Japan Earthquake,
followed by the tsunami, and then the Fukushima nuclear accident—it wouldn't be surprising that you might pick up this book,
flip through it, and perhaps take it home or click *buy*. I hope so.

But if you are none of those, why might you be interested in the
story of a disaster that occurred years ago at a remote nuclear facility? Because it's an incredible story in and of itself. Even though I
lived it, I could barely believe the scenes I witnessed.

The crisis leadership—the courageous, often selfless initiative—
that emerged from a group of those involved is also a big part of
the story. The leadership lessons to be gleaned are of enormous
value on so many planes.

Fifty minutes after the 9.0 earthquake hit the nuclear plant
known as Fukushima Daiichi, a tsunami 45 feet high engulfed the
facility, knocking out electrical power and all the reactors' safety
systems. These circumstances led to significant core damage in
three of six reactors. Buildings exploded, unleashing unknown levels of radiation throughout the countryside. Operators faced

countless dangers; two of them died while many others, even though fearing for their lives, carried out life-or-death missions to vent the reactor containments. There were no lights, no controls, no reason to stay, and every reason to flee. One operator said later, "Three times, I thought that I would die."

Their boss, Ikuo Izawa, knew he had to guide them and give them a reason to stay through the challenges they faced. How does a leader lead in those conditions? How did Izawa get his workers to follow when their emotions and their common sense told them to leave? During those cataclysmic days, Izawa and others succeeded as leaders under the toughest possible conditions. The leadership lessons to be learned from these incredible experiences are invaluable—and I have set out to tell the story with this topmost in mind.

This is a story of incredible heroism under desperate conditions that could have had catastrophic consequences. You will read about amazing feats carried out by leaders, individuals, and teams. You will read about tragedy on an unfathomable scale, about people's lives ripped apart by disaster. You will read moving personal stories of courage, heartbreak, and hope.

What I have chosen to emphasize about these stories is the tightrope that leaders must walk to guide their organizations through crisis. You will see people overcome the primal urge to flee from death and instead walk a tightrope that would determine the fate of their country. You will read about the very specific ways that Japanese culture came into play throughout an unprecedented crisis, and how American support affected the recovery for better or worse. And finally, you'll read about a country coming back to life after the unimaginable occurred, grappling with a "new normal" and pondering what actions and policies might prevent such a thing from happening ever again.

Think about a time when you met with resistance to your ideas, plans, thoughts, demands, or leadership. How did you get your unwilling followers to follow you? You are unlikely to experience an accident on the scale of the Fukushima Daiichi meltdowns, where the performance of your team may mean life or death. But you probably face smaller crises regularly, from lackluster performance to mergers to downsizing. By examining leadership *in extremis*, we open a window into human nature under pressure and gain valuable insights about effective leadership on a day-to-day basis. Leadership concepts that you grapple with all the time, such as trust, defiance, fear, and followership, all came into play at Fukushima Daiichi, where a few heroic leaders responded to the circumstances as best they could. Watching them in action may help you face your next corporate crisis.

It's also a hell of a story.

WHAT HAPPENED?

It's quite possible that you have only a vague notion of what happened at Fukushima Daiichi, so here's a brief overview. On March 11, 2011, the Great East Japan Earthquake struck the Tohoku region. The event is known as "3/11" in Japan, and everyone there remembers the many lives lost and the heroes who emerged much as we remember our own 9/11. The 9.0 earthquake was felt as far away as Antarctica. The Earth moved on its axis.

The ensuing tsunami was enormous, with multiple ocean waves as high as four- and five-story buildings. It overwhelmed villages up and down the Sendai Coast. It changed the region's landscape permanently by destroying roads, bridges, homes, and buildings. It swept tens of thousands of people and other creatures to their deaths in minutes.

Soon, Fukushima Daiichi—as well as its sister plant six miles south, Fukushima Daini, and other nuclear power plants—were without vital electrical power or cooling water needed to prevent a nuclear accident. Without these vital utilities, the reactors and spent-fuel pools would begin to melt down. Most of the emergency equipment, facilities, and first responders in the area had been wiped out by the earthquake and tsunami. It was imperative for the world to help Japan rescue the living, recover the lost, minimize nuclear meltdowns, and repair the heavily damaged infrastructure.

Hundreds of square miles of contaminated land were rendered uninhabitable for years to come. Eventually, approximately 160,000 people would evacuate from around the crippled nuclear plants.

Among those in danger were about 153,000 American expatriates, plus innumerable tourists.

Understandably, all this damage knocked Japan to its knees. Fear coursed throughout Japan and the world. But heroes emerged in those days. Untold numbers of Japanese rose to the challenge, as well as thousands of Americans after President Obama authorized Operation Tomodachi (friendship)—the first American military operation to have a foreign name—to help rescue, recover, and supply the people of Japan. Twenty U.S. ships; 140 aircraft; 19,703 military personnel; 160 search-and-rescue missions; and National Guard troops from Mississippi, Alabama, Kentucky, and Guam responded. Cub Scout packs, church groups, and schoolchildren gathered contributions for the victims.

While the ocean had viciously risen up against the Japanese, it soon brought relief. As they looked toward the east where the destruction had come from, another enormous force, this one of human origin, headed their way, led by the USS *Ronald Reagan*, flagship of the Navy's Carrier Strike Group 5. Sprayed by radioactive fallout in the process and despite the risk, the *Ronald Reagan*'s air crews continued their work, while maintaining readiness for the U.S.-Japan alliance.[1] U.S. troops took sole responsibility for reopening the Sendai Airport and accomplished the feat in a remarkably short time. Meanwhile, Japan was rising from its knees.

Owned by the Tokyo Electric Power Company (TEPCO), Fukushima Daiichi had six reactors and seven spent-fuel pools, while Fukushima Daini had four reactors and four spent-fuel pools. (In Japanese, *daiichi* means "number one" and *daini* means "number two.")

At Daiichi, three reactors were operating at the time of the accident (Units 1, 2, and 3), while the other three (Units 4, 5, and 6) were shut down for maintenance. In the end, all three operating reactors experienced near-total core damage. Due to the reactor meltdowns, a massive amount of hydrogen gas was released into the reactor buildings, setting up the potential for big explosions. As stunned viewers around the world watched on television, the Unit 1 reactor building exploded on Saturday, March 12. The Unit 3 reactor building exploded on Monday, March 14. Unit 2 experienced total core damage, but an explosion was averted because the Unit 3 explosion had blown a hole in the side of the building that let the hydrogen gas escape. Unit 4—though not in operation—inexplicably exploded the next day.

The geography at Fukushima Daini helped tremendously. At Daiichi, geographical conditions required that the plant be built at a lower elevation than Daini. During construction, 82 feet of earth were excavated at the Daiichi site. The higher elevation at Daini limited the damage from the tsunami. As a result, two power sources remained energized. Nevertheless, the operators at Daini had to work heroically to save the reactors from a similar fate—and somehow, they succeeded.

In addition to the human cost, the economic impact of the earthquake, tsunami, and resulting nuclear accident cost Japan more than 2.5 percent of its GDP. It was the strongest earthquake in Japan since the Jogan earthquake of 869,[2] and the costliest natural disaster in recorded history at about $300 billion. (Factoring in other indirect economic losses, the cost may ultimately have exceeded $345 billion.) The death toll stands at 15,870, while another 2,846 people are still listed as missing. More than 340,000 people in the tsunami zone had to relocate to temporary homes. The tsunami at Sendai surged six miles inland, leveled 130,000 buildings, and damaged more than 746,000 structures.

HEROISM AND *TOMODACHI*

As the disaster unfolded, unlikely heroes emerged. Taylor Anderson and Monty Dickson were two of them. Americans, both in their early twenties, they were teaching English in Japanese elementary schools. Taylor was from Virginia and Monty from Alaska. Monty's parents died when he was young. His sister, Shelly, raised him. Taylor and Monty shared a deep commitment to their students. When the tsunami warning sounded, they ushered the children to safety, then proceeded to help others. Monty sheltered on the third floor of a school administration building. Taylor, after making sure her charges were safe, left the tsunami shelter on her bicycle to find others to help. When the waves came, both young teachers perished.

A few years after 3/11, I talked to Taylor's sister and expressed my condolences and appreciation for Taylor's work and heroism. She said, "Taylor's not gone." I thought for a moment that she was in denial about her loss, but she continued, "She's just away from us. Taylor's a volunteer in heaven right now."

In times of disaster, people come together. America's Operation Tomodachi was aptly named, because it is in our nature to help our friends, and it is part of our country's mandate to help where help is needed. We deploy our military around the world and dispatch our nuclear experts to provide their wisdom without regard to politics. America is not always perfect, but Americans are compassionate.

The story I tell in this book is one of cooperation as well as leadership, and I am proud to have actively participated in the effort to recover from one of the worst disasters of our time.

CHAPTER 1

THE MAKING OF A CRISIS LEADER

*It took me years to realize that my emergency management
education started in my youth.*

As you must have gathered by now, I had a unique role to play
during the Fukushima nuclear disaster. For me, it started on
Tuesday, March 8, 2011, three days before the earthquake. I was
sitting in a Regulatory Information Conference (RIC) in
Washington, an annual gathering of leaders in the nuclear industry, to discuss the state of things and learn from one another. As I
listened to one of the speakers, my mind drifted. I thought about
the fact that I wouldn't be at the following year's conference because I was about to retire. I reflected on my career and felt proud
of what I'd achieved but wondered if I could have done more. I
couldn't have known that my life would change dramatically just a
few days later, and that I'd have many opportunities to prove myself
further in a role that I'd unknowingly been preparing for my whole
life—a role that I and my team would play for a full year.

My father had been an operator at a large oil refinery and told
me stories of problems he'd encountered there, including fires,
component failures, and many other challenges that arise when
operating a large refinery. For decades, he was also a volunteer
firefighter. We lived on the Ohio River. I watched as he recovered
the bodies of children and others who had drowned in the river.

I watched as the firefighters doused building and car fires. It took me years to realize that my emergency management education started in my youth.

As a young man, I served five years in the Air Force as an explosive ordinance disposal technician. In that job, I learned a great deal about risks. My biggest challenge in those years was responding to the second uprising at Wounded Knee on the Pine Ridge Indian Reservation. Detonation of several homemade bombs occurred after the killing of two FBI agents on June 26, 1975. Tensions were extremely high on the reservation, and during the uprising, we had to clear the town of any unexploded improvised explosive devices. Because they'd lost colleagues, the FBI responded in force—in fact, it was the largest assembly of FBI agents in history. Their presence raised tensions further. I remember traveling to Pine Ridge in an FBI Ford LTD, which was the most obvious vehicle on the reservation. As the agent drove 90 miles per hour down narrow roads, locals pulled out in front of the car, attempting to run it off the road. At one point, the agent pitched a handgun into the back seat where I was sitting and said, "Here, you may need this!" I immediately tossed it back to him and said that I needed him to protect me.

BROWNS FERRY

In 1978, after leaving the Air Force, I started my commercial nuclear career in nuclear construction but soon found myself in reactor-operator training. I studied the ins and outs of how the plants worked and how to operate them. I became an entry-level operator at the Tennessee Valley Authority's Browns Ferry Nuclear Plant, a three-reactor plant in Northern Alabama. I couldn't have known it at the time, but that first job was fateful: The Browns Ferry reactors were almost identical to those at Fukushima Daiichi and Daini.

Four years before my arrival at Browns Ferry, a fire broke out in the Unit 1 reactor building, engulfing a containment building and disabling the emergency core-cooling systems. Workers had caused the accident by using an open, lit candle to look for air leaks in a reactor containment penetration seal. The seal caught fire, and the fire spread rapidly. The loss of key electrical systems from the fire resulted in steam pressure building up in the reactor—much as it would at Fukushima Daiichi more than three decades later.

In the control room, the control board lights were blowing up in their sockets as the fire caused low-voltage systems to mix with high-voltage systems. This threatened the operators with electrocution if they touched the control boards. Eventually, the lights simply went out. There were no functioning normal or backup water systems remaining. Thus, reactor core cooling was impossible. As the operators tried desperately to control the safety systems manually, smoke filled the reactor building.

One operator showed ingenuity. He remembered an alternate way to open key valves that would enable reactor core cooling.

Crisis averted! And, to take care of the fire, another level-headed operator took matters into his own hands, tying a fire hose to some scaffolding and pointing it at the fire. The fire was out in a matter of minutes. At Fukushima thirty-six years and eleven days later, with no electrical power and no water injection systems available, another heroic operator would remember an alternate way to connect fire trucks and pump water into the reactors.

The heroic individuals who saved Browns Ferry were my mentors. Neither they nor I understood how important their wisdom and leadership were until I faced the challenges of Fukushima.

I worked the entry-level operator position at Browns Ferry for a few years and enjoyed manipulating the valves and controls in the field to control the reactors. My mentors taught me details about the plant that went far beyond what most engineers and operators know. We held contests to see who could name the manufacturer of the most innocuous valves in the plant, just to stump the rookie! Those contests were a big part of my growth as an operator.

After a few years in my entry-level position at Browns Ferry, I trained for and passed an initial operator examination administered by the Nuclear Regulatory Commission (NRC). During my time as a reactor operator, the NRC initiated a new type of requalification examination for operators. Many of my mentors—the ones who had saved the reactor—failed the exam. My supervisors tasked me with conducting remediation training for my mentors: It was my turn to help *them*.

A few years later, I took an operator-training instructor position at Carolina Power and Light's Brunswick plant, a two-reactor plant almost identical to Fukushima Daiichi. There, I passed an NRC Senior Reactor Operator Instructor Certification examination. A few years after that, I went to work for the NRC as an operator licensing examiner. NRC examiners write and administer those government exams that my mentors had failed, as well as initial exams for an operator license. The examiner job exposed me to many types of operators and reactors and strengthened my knowledge of the foundations of plant operations and how operators respond to accidents—especially the Fukushima design.

THE NUCLEAR REGULATORY COMMISSION

The NRC is responsible for ensuring safety in our nuclear plants. The U.S. president appoints its five commissioners, who are then confirmed by the Senate for five-year terms. The president also picks the chairman, but it is Congress and not the executive branch that oversees this agency. During my early NRC career, I served as an assistant to the executive director of operations and NRC chairman. Perhaps most significant, I was selected for a fellowship as a legislative assistant to a U.S. senator, working in his office on the Hill. This assignment was key in my education as a diplomat and another crucial building block in my training for Fukushima. These positions exposed me to policy, diplomacy, and strategic issues, and I learned how to deal with senior officials as well as politicians.

My first international nuclear crisis experience came in 2003. I was asked to join an International Atomic Energy Agency expert mission at the Paks reactor in Hungary, after an accident in a spent nuclear fuel pool caused the release of a small amount of radiation. The public was outraged, and the Hungarian government was compelled to seek out independent nuclear experts to review the causes and consequences of the accident.

The night we arrived in Hungary, we had dinner with the plant superintendent. He told me that, in the aftermath of the accident, the people of Hungary believed him to be incompetent or a liar or both. This perception struck me as significant. I immediately recognized that when you become a sinner in the public eye, the best way to become a reformed sinner is to borrow the credibility of someone else until you regain your own. The Hungarians very much needed the credibility of international experts at that moment.

The most significant leadership lesson from Paks was our discovery that production pressure had forced the utility to use supervisors to conduct independent safety reviews for a new plant component. Supervisors are always available and can be forced to work longer hours. Once the supervisors had completed their (inadequate) safety reviews, they turned their findings over to their staff for the safety review process. Of course, the subordinates assumed that their bosses had done the review correctly, and they passed the analysis straight through. No matter what the industry's record is or how routine a task might be, breakdowns in safety can still occur suddenly and without notice.

This was another experience vital to my readiness for the Fukushima assignment. Paks exemplified the value of independent assessment in penetrating the "myth of safety," as I'll explain later. As I progressed in my career, each assignment I took on helped me hone my skills.

Nearing retirement, as I reflected on my career, I felt prepared for the Japan assignment. I departed the RIC on March 10, 2011, thinking that soon I would announce my departure and fade off into the sunset.

That didn't happen. What happened instead makes up the bulk of this book.

Why a book and why now? I have contemplated writing a book ever since the Fukushima accident, but two activities kept me busy: my consulting business and the completion of my doctorate. During my doctoral work, I traveled back to Japan to interview the operators and their leaders who were involved in the accident. As I heard the operators' heroic stories, I was moved all over again.

Many of them suffered from post-traumatic stress disorder and still became overwhelmed with emotion as they spoke of their experiences. One told me that they all felt like American Vietnam veterans. They believed they had fought for a noble cause, for which some had given their lives, yet their countrymen continued to think of them as villains. They didn't feel they could tell their stories in Japan.

Those operators knew me from the time I had spent with them. They insisted that they wanted me to tell their stories—not as an official account or the reportage of investigators or the press, but their real-life, behind-the-scenes stories. That heartfelt plea, more than anything else, was my reason for undertaking this project. And, beyond the experiences of these individuals, I wanted to write about the role of the U.S. nuclear experts and embassy staff after Fukushima. I want to help leaders learn from the heroic acts of the Fukushima operators.

CHAPTER 2

EXTREME-CRISIS LEADERSHIP

*In an extreme crisis, leaders may face imminent death,
but they must cement their feet.*

There is a unique category of leadership known as *extreme-crisis leadership.* An *extreme crisis* is defined as "a discrete episode or occurrence that may result in a great and intolerable magnitude of physical, psychological, or material consequences to or in close physical or psycho-social proximity to organization members."[1]

An extreme crisis is one that is beyond the fathomable, where conditions threaten the physical welfare of the leader(s) and followers themselves. We often think about this in a military context or about first responders. Yes, leaders in those situations are extreme leaders, but in those contexts, death is an acknowledged risk. The military considers this potential for loss of life in their training, housing, and strategy for developing those war-fighters.

But in most businesses, we do not go to work expecting to face death.

I became interested specifically in extreme crisis within a business context—a context that does not usually prepare its participants for such a thing. Most businesses prepare for such things as fire, flood, or an active shooter. In those crises, the strategy is usually to remove leaders and workers alike from the crisis as quickly as possible. In an extreme crisis, however, leaders may face imminent death, but they must cement their feet.

In my doctoral dissertation[2] I examined several extreme events to better understand the role they can play in expanding our existing knowledge of crisis leadership. I interviewed those who played direct leadership roles during extreme events such as the Fukushima Daiichi nuclear reactor explosions, the Fukushima Daini response, the Deepwater Horizon oil rig explosion, the Three Mile Island nuclear accident, Superstorm Sandy, and other big events. I spoke with members of the Red Cross; U.S. and Japanese military officers; and the president and chief executive officer of the Institute for Nuclear Power Operations, Admiral Robert Willard. I also interviewed individuals involved in national leadership, for example, those who have played roles in the White House Situation Room.

My findings confirmed the research[3] about extreme-crisis leadership—and went beyond. Those studies concluded that the performance of leaders is second only to the cause of the event itself in determining the outcome of a disaster. In my research, I found that different leadership skill sets are needed for different types of disasters. Each disaster brings specific demands for extreme-crisis leadership. Ultimately, there is no one unified theory of extreme-crisis leadership. In this book, you'll read about leaders whose specific personal skills made a difference in the outcome of the Fukushima accident. This book is about the extreme-crisis leaders of 3/11.

Let's begin by reviewing the heart-wrenching stories of leaders like Taylor Anderson and Monty Dickson, the young educators in the Tohoku region I mentioned earlier, along with educators, principals, and students in Sendai. As I shared, Monty and Taylor were both English-language teachers in the Japan and English Teaching Program (JET; participants were also individually called JETs), both saved students, and both lost their lives horrifically, swept away by the tsunami. I consider them to be two of the heroes of that terrible day, though many other educators—principals, teachers, administrators—rose to the unexpected challenge and joined their ranks, saving children and helping others. Even some of the children became extreme-crisis leaders.

In Japan, it is common for schools to practice emergency procedures for earthquakes and tsunamis. But, for the Great East Japan Earthquake and tsunami, the planning had not been sufficient. This quake and subsequent waves were much bigger than anyone could have imagined, and the waves traveled much farther inland than anyone could have predicted. Schools were caught unprepared and many individuals perished.

People who used some of the extreme-crisis leadership skills discussed herein performed heroic, almost miraculous, feats. I've read two books on how school administrators and students reacted to the events of 3/11: *The Power of the Sea* by Bruce Parker (Palgrave Macmillan, St. Martin Press, NY, 2012) and *Ghosts of the Tsunami* by Richard Lloyd Parry (MCD-Farrar, Straus and Giroux, NY, 2017). They provide extensive insight into the destruction in the Sendai area, along with detailed accounts of the response of the school districts.

The earthquake and tsunami struck during the dismissal of children from school Friday afternoon. Parents were already on the road to pick up their children, children were eager to get home, and principals and teachers were ready for the workweek to end. Within minutes, a sunny day like any other was thrust into an extreme crisis.

Kama Elementary School—a mile from Ishinomaki Bay and near two rivers that connected to the sea—was far enough away that it was thought to be safe from tsunamis. In fact, it was a designated tsunami shelter. During the tsunami warning, nearby residents and parents started heading there for safety. After the tsunami inundated the city's seawall, it began its attack on homes. The rivers and canals worked as funnels for the water to flow farther inland, much farther than anyone expected. At Kama, as the tsunami approached faster than anyone could judge, the children were ushered to the roof of the school. Teachers directed arriving parents to the roof as well, as the monster waves sped toward them, taking everything in their path. As a wave violently struck the building and then receded, parents and teachers remaining on the first floor were pulled out. Staff attempted to throw fire hoses out the windows to them, but most of them—and most still in their cars or on foot—were swept away. Afterward, some thirty students sat patiently in a third-grade classroom waiting for their parents to arrive. No one came. Eventually, friends or relatives were summoned to pick them up and explain the inexplicable. Kama had been assumed to be a safe place, but it proved otherwise. As we learned on a grander scale at Fukushima, if you don't believe something will happen, you aren't prepared for it.

Near Kama, at Kadonowaki Elementary School, teachers led the children to the peak of a 180-foot hill. The tsunami wave surrounded the peak. Fires raged and smoke billowed all around them as the children witnessed cars crashing into the building and hundreds drowning as they attempted to reach the peak. Another school, Nobiru Elementary, was near a canal that connected to the sea at both ends. It had an earthquake-reinforced gym across a courtyard from the main building. After the earthquake subsided, the principal moved all the students to the gym. She chose not to lead them up a nearby mountain path, fearful that trees would fall on them. Neighbors, including many elderly, joined them in the gym. As the tsunami approached, people screamed about how fast it was moving. The principal then decided that the main building would be safer and tried to move the students and residents there, but it was too late. They ran back into the gym, and soon the cold, black water slammed into them. As the students huddled on the stage, water filled the building. Only a small second-story balcony was safe, but there wasn't enough room for everyone. As the whirlpool of water surged throughout the building, it swept many away. Then it stopped, and soon darkness and snow fell. After hours of trying to keep warm, many of the elderly succumbed to the elements. In all, eighty-five people were lost that day at Nobiru.

One of the worst tragedies took place at Okawa Elementary School, which was located just a few hundred feet from the Kitakami River, close to Oppa Bay. The dreadful wave crushed the school, jammed trees deep into its structure, wiped out most of its walls, and left only the foundation. Of the eighty-two students who had not been picked up by parents before the wave, seventy-four perished. One student and teacher had run up a nearby hill. They survived. The other seven survivors were gathered up by the wave and swept across a seawall, but, unbelievably, they survived. The earthquake killed ten of the thirteen teachers and staff.

Flowers, messages, and memorabilia covered the site after the tragedy, and it became an informal shrine. A message from a grieving mother read that she was sorry that she couldn't protect her daughter, that her daughter would never have children. Parents blamed the school and the government, because they felt there had not been sufficient preparation for tsunamis.

Many of the bodies were not found, making the grieving worse. One father who did find his son's body waited endlessly for his daughter's to be found, to no avail. The pregnant mother of a young lost boy learned to operate a backhoe and spent weekends searching for her son, ignoring the pleas of her family to stop.

Just five miles from Okawa, on the Unosumai River, which flowed into Otsushi Bay, was Kamaishi East Junior High School and, next to it, Unosumai Elementary School. Tsunami waves advanced through the water and reached both schools. In an amazing circumstance, some 212 junior high and 350 elementary students survived. Though the intercom system was damaged and there was no way for staff to communicate with the students, the junior high schoolers followed a pre-arranged plan and escaped the building. After the earthquake struck, the elementary students watched their junior high counterparts leave the building and head upward. Without guidance, they ran out of their building, joined the older students, and all went hand in hand to the evacuation site. Once there, one student was worried that it was not high enough to protect them from the tsunami, so they headed even higher, the junior high students guiding their younger companions to safety. At one point, they ran across a group of kindergarten students whom they shepherded to higher ground as well. After a mile climb, the students watched the destruction of schools, communities, and lives below them. The terrifying scene would be with them for life, but they had somehow navigated to safety.

Throughout the school district, these scenes played out over and over. At Toni Elementary School, the principal led her students quickly out of the school and up a hillside in time to elude the wave. Overall, of the 2,926 elementary and junior high students in the Kamaishi district, only five died or went missing, and those children had not been at school that day. More than 1,300 people vanished from the city of Kamaishi, yet the number of schoolchildren lost was proportionally much smaller. The reason? I believe it was their preparation. At Kamaishi Elementary, many of the students were already at home. Some of the children remembered their school videos and drills and used that knowledge to survive. During the big quake, one middle school girl remembered that her training instructed her to help others, and she ran to her neighbor's house to rescue an elderly woman. Tragically, a chest of drawers fell on her and she died. She was one of the five lost schoolchildren of Kamaishi.

Before the big quake, Kamaishi city had been concerned about preparedness, but a program for tsunami training went mostly unheeded by residents. A leading professor from Gunma University, Dr. Toshitaka Katada, had tried to teach adult preparedness classes without much success. He came to believe that if he could teach the children, perhaps over time the city would be more prepared. As a disaster social-engineering expert, he had studied past destruction by earthquakes and tsunamis. He integrated a life-saving model into every aspect of the school's curriculum, including math classes, where kids studied the speed of a tsunami. Science classes taught the physics of tsunamis, and history classes studied their history in the area. Dr. Katada's program also focused on long-term survival and helping others. The students learned first aid, even how to run a soup kitchen.

In tsunami areas, prior to Dr. Katada's mantra, people often followed the philosophy of *tendenko*.[4] They would say *tsunami tendenko*, which means *flee separately*. Dr. Katada modified *tendenko*. Like the rule for a fire of "stop, drop, and roll," he taught the students three steps. First, don't go inside—run to high ground. Second, don't rely solely on planned evacuation routes because the damage is unpredictable. And third, help others who need it. Dr. Katada taught his students how to be crisis leaders themselves and not wait for leaders to guide them to safety.

There had long been a policy of reuniting schoolchildren with parents during disasters. A panel of experts recommended a reversal of this policy in 1996, but it was not officially changed.

Unfortunately, even though they knew that students and administrators had been trained to survive, parents couldn't stop themselves from rushing to the school. In one example, within a cluster of thirty-three schools, 115 children picked up by parents died while the children who remained did not. This lack of trust caused the deaths of many parents and their children.

The process of using buses to send children home led to yet more deaths. A bus driver decided to wait for the earthquake to subside before dropping off seven children. When he saw the wave coming, he tried to rush back to the school but the wave overtook the bus and they all died.

Overall, although most districts had not had the benefit of Dr. Katada's guidance, a disproportionate number of students survived because they had at least some training in how to handle themselves during a tsunami and earthquake. Even in schools that suffered many deaths, the students were safer overall than the general population of the towns.

Though each town in the area had a high seawall to lessen the blow of a tsunami, the waves were too much for these structures. Many evacuation sites were simply not high enough to protect the population.

In cold, practical terms, tsunami walls are expensive. Builders must make judgments about acceptable levels of risk and the potential consequences. These are the same types of calculations that designers make when building nuclear plants; every industry has some version of the need to balance cost and risk. In this process, they can easily underestimate or misjudge the risks. That is why emergency planning is so necessary—especially in places where systems are insufficient to guard against potential risks. Emergency planning won't always be enough; it wasn't for the schools and nuclear plants in Fukushima Prefecture. At that point, extreme-crisis leadership concepts take hold. Leaders and those affected must use *sensemaking*[5] and *teamsense* (my word for sharing sensemaking abilities among an entire team) to make good decisions and act on the facts they know, such as the speed and magnitude of the wave.

We can see in the stories of the schoolchildren that emotions play a significant role in extreme-crisis decision-making. Whether it's parents who do not trust that their children can survive without them or administrators who misjudge the gravity of the situation, emotions matter hugely in an extreme crisis. Taylor Anderson and Monty Dickson were two heroes among many, and again, we honor them. The best way to do that is to spend our energy on eliminating the need for heroes.

CHAPTER 3

OVERVIEW

"This airline doesn't give a sh@t about us,"
one flight attendant spat. Rather than argue with her, I
just said, "Well, you may be right, but I know
one thing that they do give a sh@t about,
and that's this aircraft."

On March 11, 2011, conditions at Daiichi and Daini were extremely dire. In fact, from March 11 to December 21, 2011, conditions at the nuclear plants remained tenuous, as aftershocks and tsunamis continued to occur at regular intervals (hundreds a day). Immediately after the initial tsunami, Daiichi plant superintendent Masuo Yoshida had to make crucial decisions about how to deal with the reactors, including whether to introduce ocean water into them to stop the meltdowns, and whether to vent radioactivity into the environment, thus contaminating large sections of Japan. Yoshida faced unfathomable challenges. He understood that the order he received from the company's president—not to use seawater—was antithetical to safely cooling the reactors. His boss was worried about the company itself and knew that seawater would destroy the reactor. Other Japanese leaders worried about seawater causing a "re-criticality" (uncontrolled restart) of the reactors.

Yoshida had more humanitarian concerns. His decision might very well involve an act of leadership defiance. Should he put seawater in the reactor, stop the meltdown, prevent a radioactive plume, and potentially save many lives? Or should he listen to the president of the company and the prime minister, those who were telling him not to do it? That's a big decision for anyone to make, particularly under that day's intense pressure. Buildings were exploding, radiation was escaping, it was dark, and he was afraid and worried.

Put yourself in Yoshida's position. You know that introducing the water is the right thing to do, but you're being told by the president of your company not to do it. What would you do? Yoshida continued the seawater injection. He knew it was the right thing to do. Decision-making often involves moral judgment. Decision-making may mean defying someone who is asking you to do something for the wrong or improper reasons. As a leader or worker, these can be the toughest decisions you face.

HELP FROM THE STATES

In the wake of the nuclear plant disaster, Japanese Prime Minister Naoto Kan called U.S. President Barack Obama for help with the nuclear catastrophe, and Obama promptly expanded Operation Tomodachi to include support for the response. In total, the U.S. government dispatched some 150 responders, sixteen of them from the NRC, to Japan (with thousands supporting them back in the States) to provide equipment and advice on controlling the reactors, spent-fuel pools, and other protective activities. The U.S. Navy floated barges[1] containing fresh water for the reactors. It designed and built a freshwater pumping system and flew it by U.S. Air Force C-17 crews from Australia to Japan. The United States supplied robots, helicopters, radiation suits, radiation detection equipment, and crucial water-pumping apparatus. In all, it was an enormous supply effort.

I had returned from the NRC's RIC conference and was in Atlanta on the day the earthquake and tsunami attacked Japan. I watched the devastation occur on television at home. A friend called me while I was at a Walmart getting gas in my pickup truck. "Chuck," he said excitedly, "what's happening with this nuclear plant in Japan?"

I was confident that the operators at Daiichi would soon restore electrical power, enabling the reactors to be cooled. "It's going to be OK," I said. "Everything'll be fine. They're going to solve the problem—don't worry about it." How wrong was I? As I continued to watch events unfold in Japan throughout that weekend, I kept thinking, *Soon, they will restore emergency power; they'll stop the accident from happening.* I knew the clock was ticking. Without restoring power and water quickly, major nuclear fuel damage would occur.

As the hours ticked by, I grew frustrated at the inaction—but I knew that more must be going on than we could see on television. I thought about those heroic operators at Browns Ferry, thirty-seven years earlier, desperately trying to extinguish the fire and inject water into the reactor. I imagined that the Fukushima Daiichi operators were frantically working and would somehow restore power soon.

Then the Unit 1 reactor building exploded. At that point, I knew for certain there was serious nuclear fuel damage. I knew that hydrogen could be released, and that there might be small hydrogen fires as a result. Even with my extensive BWR (Boiling Water Reactor) experience, however, I could not have imagined a hydrogen explosion that big.

It would be the first live-stream nuclear accident in history. There had been no public pictures of the Browns Ferry fire, and at Three Mile Island, there had been only still pictures. It was days before any video emerged from Chernobyl. Despite the ensuing claims by some international representatives and the Japanese that there had been no nuclear fuel damage, I knew that couldn't be the case. The only mystery to me was what set off the explosion. Later, when my neighbor found out that I was leading the U.S. nuclear recovery effort in Japan, he said, "You? Mister Everything's-Going-to-Be-OK? *You* are going to oversee this effort?"

"Yeah, I'm the guy," I said.

DISPATCHED TO JAPAN

At the time of the disaster, I was working as a leader in a nuclear plant construction organization in Atlanta. I was not organizationally connected to the NRC's Washington Operations Center and not privy to any details of what was unfolding half a world away. After President Obama expanded Operation Tomodachi, he decided to dispatch an additional team for the nuclear event, so he called the NRC chairman. In Washington, the decision was made to send two senior reactor analysts (SRA) to Japan, and they arrived on Sunday, March 13. The NRC needed a leader for this team, and my name, among several others, was submitted to the executive director for consideration. Given my experience as an operator, as well as my political and diplomacy background, I was chosen to lead a sixteen-person team.

The call came on Monday: "You need to be at the airport in three hours." I was to report to Ambassador John Roos in Japan. We were tasked with helping him make crucial decisions about protecting American and Japanese citizens in Japan. Suddenly, I was off on a 7,000-mile journey to a country I'd never visited, to navigate a culture I didn't know, in a language I didn't speak. I was to be the senior federal executive in charge of the team.

I called my wife, Beverley, who was just leaving the school where she taught, to ask that she meet me at home to help me pack and find my passport. I changed into what I thought would be a suitable "emergency response" uniform, including a shirt with an NRC logo on it. Then she drove me to the airport.

During the drive, I talked on the phone to the NRC chairman, who conveyed my marching orders. It was clear that the U.S. government was fully behind this mission. The only constraint was that we work as closely as possible with the Japanese government authorities, rather than just the people of TEPCO. This order seriously constrained our efforts.

I made my flight, but just barely, and had to spend the night in Dallas because there hadn't been enough time in Atlanta to grab any potassium iodide (KI) for our team. KI is crucial for blocking radioactive iodine from the thyroid. I arrived in Dallas after midnight, and a member of the Dallas NRC staff brought the KI pills to my hotel room in a small cooler. It felt like an illicit drug deal.

The next morning, I boarded a flight to Tokyo. It was clear that the passengers and flight crew were concerned about flying there, what with the continuing aftershocks and unfolding nuclear disaster. I heard murmurs about radiation exposure.

Noticing my NRC logo shirt, a flight attendant asked me why I was going to Japan. After talking with her, I settled into my cozy government-provided coach seat at the back of the aircraft for the fourteen-hour flight.

Just after we reached cruising altitude, the lead flight attendant came to me and said, "Mr. Casto, would you please get your luggage and come with me?" I thought she was going to throw me off the plane for being a radioactive nuclear guy. I assured her that I hadn't caused the accident—I was going to help fix it. As it turned out, she took me to first class, where she directed me to an empty seat. Wow!

I had brought some reference material to read on the flight, and I thought, *Great, I'll read and get some sleep.* I was wrong. For hours on end, the flight crew took turns sitting with me and asking me about how reactors worked and whether they should be traveling to Japan. They even questioned me about the radiation-measuring device I was wearing. When they asked about the need to monitor flight crews for radiation exposure, I thought, *Oh no, I'll be the basis for their union action!* They'd question me for a while, step away to take care of their duties, then return with more questions. I tried to reassure them.

"This airline doesn't give a sh@t about us," one flight attendant spat. Rather than argue with her, I just said, "Well, you may be right, but I know one thing that they do give a sh@t about, and that's this aircraft." I told her that the airline would never fly one of their precious planes into a radiation plume, because it would cost too much to decontaminate it.

Bottom line: This was a small inkling of the fear swirling around the accident and around the world, and I would soon be at ground zero of that fear.

WEIGHTY DECISIONS

The next thing I knew, I was at the embassy in Tokyo. Embassy leaders would need to make some very important decisions about the protection of Americans. There were discussions about whether to evacuate the entire country; discussions about allowing TEPCO to evacuate its operators at Daiichi, thus leaving the site unoccupied. That action would be catastrophic. We talked about the fact that stopping the accident might result in exposure to lethal doses of radiation. As I said, these were big decisions. Just hours earlier, I'd been at an Atlanta Walmart. Now I was faced with matters of life and death in a place completely foreign to me.

Meanwhile, the heroic operators at Daiichi and Daini were desperately trying to restore cooling to the reactor cores. Out of view of the world, a tragic drama was unfolding at those two sites, where humans did battle with the forces of nature and physics on an unprecedented scale. They may not have chosen their roles, but they became heroes nonetheless.

KICKOFF OF THE NUCLEAR SUPER BOWL

We are about to take an amazing ride through the events that occurred at Fukushima Daiichi and Daini. The next few chapters focus on the leadership abilities of three key leaders. At Daiichi, Ikuo Izawa, who grew up a local Fukushima Prefecture boy and was serving as a TEPCO control room shift supervisor for Units 1 and 2. He would play a crucial role in the aftermath of the disaster. Maintenance Manager Takeyuki Inagaki guided the recovery strategies from the Emergency Response Center[2] (ERC). Masuo Yoshida was the site superintendent at Daiichi and is an important part of the Fukushima story. Many books and reports have chronicled Yoshida's leadership, however, so I have chosen to focus mainly on the two leaders who reported to him during the accident.

At Fukushima Daini, the plant six miles south of Daiichi, Site Superintendent Naohiro Masuda fought diligently to lead his staff through the earthquake and tsunami. They were operating under somewhat better circumstances, yet were perilously close to a fate similar to that of Fukushima Daiichi. Without Masuda's able leadership, that plant, too, would have suffered a horrendous fate.

There were countless heroic acts at both plants. The purpose of these chapters is to follow the efforts and decisions of these three leaders and view the accident through their actions. This book is not a detailed "tick-tock" story of the accident, nor does it collect all of the memorable stories from those desperate days. Some amount of the technical detail is summarized to help make the stories more understandable.

What is important to remember about these workers is that they were just regular people who found themselves in extreme circumstances. They were not soldiers or first responders, trained and ready for life-or-death situations. These were workers in a highly technical and, yes, high-risk industry trained for "design basis"[3] events. The situation they faced was far beyond their training and preparedness. And yet they fought on.

On the American side, we began calling the situation the "nuclear Super Bowl." It was one of the most important nuclear events ever and was watched live by millions of people around the world in real time, and the images were deeply troubling.

Amid the chaotic conditions, I set out to get our team organized. It was anything but easy. It wasn't clear how long I was going to be in Japan. My assignment ended up lasting just about a year and became the pinnacle of my career. As an extreme-crisis leader, my experiences there taught me invaluable lessons. I tend to remember those experiences as vignettes, of which the most profound will never leave me. I have attempted to capture them here as best as I remember them.

I must say that the number of serious issues we had to address every day was mindboggling. Not all of them are discussed in this book; rather, I have focused on those that illuminate extreme-crisis leadership. We faced challenges regarding agricultural contamination; the restoration of power and water at the nuclear plants; support of radiation-exposed workers; coordination with the U.S. military and the Japanese government and military; a flood of knock-off radiation detectors that were giving hundreds of false radiation readings around Japan; concerns about the accident's impact on the U.S. West Coast; freedom-of-information requests; and hundreds of demands for media interviews—among many, many other challenges.

Before I dive into the thick of it, allow me to offer a quick tutorial on the construction of the Fukushima Daiichi and Daini reactors (Figure 1). These reactors are a model called a General Electric Boiling Water Reactor (BWR), with a Mark I containment. Some of the units are slightly newer and built just a little differently, but they are functionally the same.

The innermost structure is the reactor pressure vessel, which holds the nuclear fuel and serves as one of the barriers to the release of radiation. Then there is the Primary Containment, which is the primary barrier to release of radiation. As you can see, it appears as an inverted light bulb. Under the Primary Containment is a Suppression Pool, which holds thousands of gallons of water and is used to suppress steam coming from the reactor during accidents. It also serves as a source of water for pumps to inject water into the reactor.

During the Fukushima Daiichi accident, there were two key strategies in play. One was meant to reduce the pressure in the reactor vessel ("depressurize" or "vent"), which would allow large, low-pressure water pumps to flood the reactor vessel. For reactor depressurization, safety-relief valves (SRVs) are opened, releasing steam to the Suppression Pool where the steam is condensed. During this accident, heavily contaminated steam and water would have to be contained by the Primary Containment (drywell). Failure of the Primary Containment could lead to significant radiation release.

Therefore, the second key accident strategy became the effort to "vent," that is, relieve steam pressure, from the Primary Containment. Venting to the atmosphere, through radiation capturing filters, is a strategy that keeps the Primary Containment from exceeding the design pressure and causing a major physical rupture of the containment. Venting can release a small amount of radioactivity, but that is much better than an explosion or rupture of the Primary Containment. It also allows the use of high-volume, lower-pressure pumps to fill the reactor cores. The design allows for venting from either the drywell or the Suppression Pool. Venting from the Suppression Pool is the most effective means,

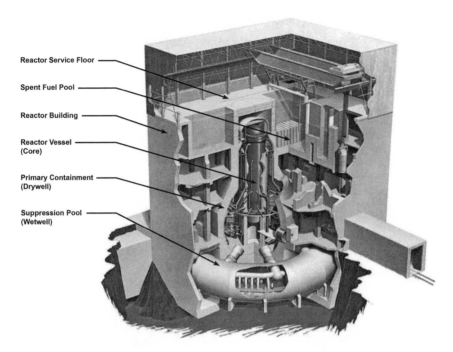

Reactor Service Floor

Spent Fuel Pool

Reactor Building

Reactor Vessel
(Core)

Primary Containment
(Drywell)

Suppression Pool
(Wetwell)

Figure 1. Design of Boiling Water Reactor

Source: World Nuclear Association, "Fukushima Accident." Accessed at http://
www.world-nuclear.org/information-library/safety-and-security/safety-of-plants/
fukushima-accident.aspx.

because that steam/air is "filtered" by the water in the Suppression
Pool before releasing to the atmosphere. Venting from the Primary
Containment does not allow significant filtering, so, obviously,
Suppression-Pool venting is preferred.

Next are the buildings that house the reactor: the reactor building, which serves as a secondary containment barrier to radiation
release; the reactor service floor used to refuel the reactor; and the
nuclear fuel storage area, called the spent-fuel pool. Ultimately,
during the accident, cracks occurred in the Primary Containment
that allowed hydrogen to escape. The reactor service floor is where
the hydrogen collected and underwent explosions.

Figures 2 and 3 show the damage to Fukushima Daiichi Units
1–4 after the earthquake, tsunami, and hydrogen explosions.
Figure 3 depicts the Unit 4 reactor, which was not operating at the

time of the disaster but sustained a sympathetic explosion related to Unit 3. The condition of its spent-fuel pool became of major concern.

Figure 2. Daiichi site damage

Source: Tokyo Electric Power Company (http://photo.tepco.co.jp/library/110316/110316_1f_chijou_2.jpg).

Figure 3. Daiichi Unit 4 reactor damage

Source: Mandatory Credit Photo by Air Photo Service (http://pinktentacle.com/images/11/fukushima_4.jpg).

CHAPTER 4

MARCH 11, 2:46 P.M., THE GREAT EAST JAPAN EARTHQUAKE STRIKES

"If I were cornered, I'd try to get out of the corner not by skill but by spirit."

—Iκυο Izawa, CONTROL ROOM SHIFT SUPERVISOR

These words come from a control room shift supervisor at Fukushima Daiichi who experienced forces of physics and nature far beyond those experienced by most leaders. How does a person lead in those conditions? Ikuo Izawa knew from the force of the earthquake—the largest he'd ever experienced—that he'd have to scram (shut down) the nuclear reactors safely. On top of that, he'd need to ensure the safety of the operators and subcontractors working on-site. In the central control room, he could see the status of the reactor, but he had no idea what danger the workers faced.

In the central control room, alerts went off, lights flickered, and fire alarms sounded. The earthquake caused a white cloud of concrete dust to fill the control room. Izawa needed the operators to stay calm, but they could barely hear one another over the racket. Some of them were beginning to unravel. Those at the control panels, thrown to their knees by the quake, began checking the status of the plant's automatic shutdown. They were looking for signs that the control rods were completely in the reactor core to shut down the chain reaction. They were reporting to Izawa with hand signals, fearing he couldn't hear them over the racket of the alarms.

As soon as they could stand up, the operators started to walk toward the control panels to work, but Izawa stopped them and told them to calm down, "Do as you were trained in the emergency drills," he said. "Check all the alerts without missing any of them; follow the instructions very carefully and confirm everything that needs to be checked. We are still having some shaking, so for now, think calmly about how to deal with this—don't rush anything. First off, don't operate the plant. Don't do anything until the quaking calms down."

As his operators began to recover from their shock, Izawa said: "The reactor has properly scrammed [shut down, fission process stopped], and the water supply has started up. We also have electrical power, so calm down and do things the way we practiced. Go!"

That got them moving. The operators were keyed up, talking quickly. When Izawa saw that, he told them, "Speak slowly!" He would repeat that many times—sometimes talking a little too quickly himself. It was hard to stay calm, even for the most experienced operators. After they had completed the emergency procedures he had reminded them of, Izawa let his staff go back to the control panels to verify the reactor conditions. Then, Izawa and his team headed for the Unit 1 reactor control panel. Izawa stood in the middle of a group of in-plant operators. Reports and notices were coming in, and he wanted to make sure that they understood the warnings. "Are you doing OK?" he would ask one after another. "Your hands aren't shaking or anything?"

It had been a very big earthquake, so Izawa knew beyond a doubt that there would be a tsunami. Their procedures stated that they must prepare to evacuate nonessential staff in the event of a big earthquake. Also, Izawa emphatically ordered the in-plant operating staff to seek safety. According to protocol, after a reactor scram, some staff are tasked with going into the reactor plant to check the status of equipment, but the earthquake had been so enormous that Izawa wasn't sure of the plant conditions. He was reluctant to send the in-plant operating staff to check each location.

Fire alarms were still sounding. Rules prevented them from turning off the alarms until they checked each location for fire, but Izawa decided that, to restore a sense of calm, he would break this rule. He reset the alarms, presuming that if they went off again, it would be because of a real fire somewhere on the site.

His top priority was the safety of his workers. He informed them that no one would be dispatched anywhere on the site without his permission. From the central control room, he repeatedly announced that nonessential workers should evacuate. As he was not willing to dispatch his operators to help guide workers out, he could only pray that they would be able to evacuate safely, as they had practiced in emergency drills.

He kept asking himself if he was panicking. At one point, he stepped behind a pillar, out of view of his staff, to collect his thoughts. He checked his pulse, monitored his breathing. Were his palms sweating? Izawa worked hard to organize his thoughts in response to the reports he was getting from his subordinates.

In the immediate aftermath of the quake, there was confusion in the Emergency Response Center (ERC). Maintenance Manager Takeyuki Inagaki understood instantly that he and Masuo Yoshida, plant superintendent, would have to lead their team through it. Their actions over the ensuing hours and days—their attempts to protect, organize, and mobilize their people—provide a remarkable firsthand lesson in leading through chaos.

It would fall to Inagaki to develop and execute strategies to address some of the most challenging conditions ever to face a nuclear plant. It was likely that his decisions would lead to success or to further damage of the reactor cores. His response to setbacks must not diminish his strategic focus. When I interviewed Inagaki in 2013, he reflected, "I thought, at least three times, 'I will die soon.'"

There was an all-hands meeting scheduled for early Friday afternoon. Inagaki was standing in front of his desk in the administration building, chatting with coworkers and waiting for the meeting to begin, when he heard a growling sound. A kind of rumble. Before he could even wonder what it might be, the jolting began.

"Everyone under their desks," he shouted firmly, and people began diving for cover. Inagaki himself continued to stand and watch the panel that monitored electrical output from the reactor units. The numbers seemed to be holding steady, but this was no brief earthquake. As the shocks grew in intensity, the ceiling panels started falling and Inagaki knew he wasn't safe where he was. He grabbed a hard hat and some safety gear and sought cover under a table, holding on to its legs as the jolts continued for a full three minutes. When the earthquake finally seemed to subside, debris and dust filled the room. Shocked and disoriented, no one spoke, but Inagaki's mind was racing. He quickly looked up at the electrical panel display, which indicated zero. The reactors were all shut down.

As luck would have it, the team had been through an evacuation drill just a few weeks earlier, and everyone seemed relatively calm. "OK, now let's evacuate," Inagaki announced in a level tone. This was reinforced by an announcement over the in-house speaker system: "Please evacuate to the parking lot in front of the administration building."

Obviously, the construction of the administration building was much less sturdy than the reactor buildings. "The architecture people were trying to kill us," he later joked. "They were focused on the safety of the reactor buildings and towers, but should have been equally concerned about the administration buildings. Our safety was very important!"

Once his people had evacuated to the parking lot, Inagaki checked for stragglers throughout the building, then went down himself. He asked his group managers to take a head count and make sure no one was missing or seriously injured. He noted that one woman with a handicap had been escorted down carefully and kindly by her coworkers. The entire evacuation process took twenty to thirty minutes. Once Inagaki had confirmed that all his people were safe, he moved to the ERC.

Normally, it takes some time from the start of an earthquake to the arrival of a tsunami. Izawa had communicated with the ERC during this time, but the ERC was in shambles. Its ceiling had come down during the quake, and the staff had been forced to move outside. Before Inagaki arrived at the ERC, every section chief there was trying to confirm the staff's whereabouts and condition. Once that was accomplished, they moved back into the ERC. The ERC was not in a calm and orderly condition at all.

BETWEEN THE QUAKE AND TSUNAMI—
SETTING PRIORITIES

As the aftershocks kept coming, some of the security staff and those from the general work section got into vehicles and conveyed instructions via loudspeakers. In the ERC, Inagaki found the operations people exchanging whatever information they had about each reactor unit. Units 1 through 3 had been in operation when the quake hit, while 4, 5, and 6 had been shut down for refueling. The working units had automatically shut down (scrammed) per protocol. The operators reported that they were confident that they could get the reactors safely to the cold shutdown[1] stage. Inagaki watched news reports coming in on the six or seven television monitors in the room, and all of them were warning of an impending tsunami. The clean-up control room was receiving the tsunami warnings as well. It was not expected to be too large—perhaps ten to sixteen feet—some reports said twenty-three feet. This meant the site would be safe from water encroachment. Izawa couldn't remember how high a tsunami might be after a great earthquake, but he imagined that the foundations of important pumps might get wet. He tried to come up with a plan for this, never even imagining that waves would reach the reactor and turbine buildings themselves. The oceanside pumps, used for cooling the reactors with seawater, might get wrecked, he thought, *but there's nothing we can do about that.*

Meanwhile, operators in the central control rooms were focusing on the shutdown function. Izawa was receiving reports and checking that everything was being done per protocol. From time to time, he would check the manual to make sure they were not overlooking any procedures. He wanted to make sure that everything was being done properly before the tsunami hit.

There were more tsunami warnings, but it was difficult to register them as real. Reports were saying the wave might reach thirteen feet, then twenty, then thirty-two or forty-eight. *No way, thirty-two feet?* thought Izawa. He simply could not envision a tsunami that could reach the reactor and turbine buildings.

MARCH 11, 3:27 P.M. AND 3:35 P.M.: THE WAVES ATTACK

Forty-one minutes after the earthquake started, the first tsunami wave crashed over the first defense at Daiichi: a 1.5-mile breakwater of 60,000 concrete blocks. Eight minutes later, a larger wave crashed over the last line of protection: the site's eighteen-foot seawall. The ERC was just 1300 feet from the seawall. When the largest tsunami waves hit, Izawa, in the windowless control room, had no idea what was going on around the site. Just as he was contemplating the possibilities, one of the emergency diesel generators shut down. Izawa suspected that someone had inadvertently caused the shutdown. As it turned out, this was the first symptom of the tsunami attack.

He did not know what to think when all the control room alarms and alerts stopped one by one right in front of him. After all the chaos during the earthquake, now the panel was dark and silent, and the operators could not understand what was right in front of their eyes. They'd lost all power. The control room lights were all out at Unit 2, and the emergency lights were barely functioning for Unit 1. Silence struck them all. There was another report that all the emergency diesel generators had shut down.

Izawa knew that, with a ten-foot wave, the seawater pumps would be lost. He'd prepared himself for that, but when he heard that all the emergency diesel generators had tripped, he couldn't believe it. He wasn't sure what the shutdowns meant. The situation was simply unimaginable.

They did have some undamaged emergency battery power (DC) at several reactors that might last for up to eight hours. The best instruction Izawa could give the operators was to check what they could physically monitor from the remaining gauges in the central control room. He had to report to the ERC about the status. Izawa had no idea how much damage they'd sustained, or what to think about the outages. Unit 1 lost all off-site and on-site AC and most DC battery supplies, starting a condition known as a station blackout (SBO). They had no idea if this loss would be permanent. The plant was flooded—rooms were full of seawater and sea life; the concrete hatches, bolted manholes, had just blown away dozens of feet into the sky. Louvered doors had lifted—the result seemed like an attack.

Izawa didn't know for certain yet if they were experiencing an SBO, though everyone was asking him. He sent two people to Unit 1 and two to Unit 2 for a damage assessment of the emergency diesel generators in the turbine building basement. These four workers took the stairs. When they arrived at the first floor, they heard a loud sound from a distance—an unusual roar. Something was seriously wrong. Another tsunami wave arrived. Climbing against the water, they barely outran it up the stairwells and jumped into the control room for safety, drenched. Water was coming into the buildings. The control room staff thought, *seawater? Unbelievable. Incredible.* They asked the in-plant operators to explain what they were describing. *Water flow to where?* The SBO connection started to sink into Izawa's mind.

Not even a few minutes later, the remaining emergency ceiling lights went out in Unit 1. There was now no way that operators could monitor the status of the Emergency Core Cooling Systems, which were designed to prevent core damage. Nor could they check on the water levels in the reactors.

The surveillance cameras used to monitor the plant were out of commission as well. There was one building that faced the ocean in Units 5 and 6, and those supervisors sent watchmen to look out for more tsunamis. The watchmen quickly realized that their position was too low—that the tsunami threatened their safety. They had to climb to higher ground. Other workers were running up the road, trying to reach high ground as well. The waves washed away many of the cars parked along the roads. Two workers in the Unit 4 turbine building—two young men—drowned at that moment.

Later, some would question Izawa as to why no one in Units 1 and 2 suffered a similar fate. "Just luck," Izawa would conclude. These drownings affected everyone deeply at Daiichi. By the time the tsunami hit, Izawa had dispatched staff to other buildings, or they might well have died, and he would have been the one who'd sent them to their deaths. The four operators he sent had made their own on-site judgment to flee, and that had saved their lives.

I might have killed these operators, Izawa thought. *It was their own decision to flee.*

All I have just described happened within ten minutes. When Izawa made a call to the ERC, they could not believe what they were hearing. He could tell they were in a panic when he told them that the situation was so severe that they needed to report to the government immediately. Calmly, he referred them to "Article 15 (of the Nuclear Emergency Act): loss of Emergency Core Cooling." This was an emergency, he repeated, and they had a legal obligation to report it.

"We've lost all the power supplies and all of the Emergency Core Cooling Systems?" the ERC asked in disbelief.

Izawa confirmed that, yes, the conditions of an Article 15 nuclear disaster had been met. He then called someone even higher in the ERC and was greeted with panic yet again. Izawa remembers him mumbling and muttering, then becoming speechless.

During that conversation, another operator rushed into the central control room and told them that seawater was flooding the buildings, that he'd seen ships drifting by, large tanks rotating. This shocked Izawa, who understood at that moment the impact of the tsunami—that it was the direct cause of what was happening in the plant.

They had conducted SBO exercises on the reactor training simulator, of course, but a total loss of DC battery power had not been part of the scenario. The drills would usually be "frozen" (the simulator stopped) by the instructor before it came to that. One of the operators looked back toward Izawa and said, "Isn't this where the instructors say 'freeze'?" Amid the chaos, they had a little chuckle. That's when they grasped, *This is real, and we own this situation; it's not a simulation.* What Izawa realized was that the tsunami wave had knocked out almost all electrical power to the reactors. He decided that the best guidance he could provide was to have his team focus on what they could physically monitor from the gauges that remained functional. "What is usable? What is viewable?" he kept asking them.

He told the operators in the control room, "Let's stand back and let the systems work." Normally he would have people go to the field to verify conditions, but not this time. He made a rule: The control room operators were not to dispatch anyone to the site without his permission. And if they did go, it was to be in pairs. As the situation on-site was uncertain, it would be best if two went together so they could discuss whether to go farther or come back. And he did not want too many pairs out at the same time. They had a limited number of in-plant operators and would have to send a new pair to rescue anyone who needed it. Izawa set a time limit of two hours for any pair to be out, since they had only a few battery-operated radios for communication.

Izawa also limited the access points. If his teams could not get into the plant at those points, they'd have to come back. They were to write their time of entry and exit on the whiteboard in the central control room.

ESTABLISHING THE RULES

When we read the story of Izawa, we see that he first consolidated his power, and then he freed his workers to follow their instincts. Before letting the first teams go, Izawa impressed upon them the importance of following the rules he had made. "My orders are absolute," he told them. "If you don't follow these rules, you're going to get yourselves killed. You got that?" All the operators agreed to his rules and obeyed them, though none of these protocols had been included in simulations—until 3/11 when they evacuated the plant. This surprised even Izawa himself.

The situation was still developing, and Izawa understood that if radiation levels increased at the central control room, they would have to evacuate it at some point. Because he knew that keeping the central control room manned was crucial to ensuring safe access to the site, he determined that he would be the one to stay behind. He completely gave up on the idea that new teams of people would come to replace them and was determined to be the person to make decisions about the behavior of his operators.

In the aftermath of the wave, the central control room staff struggled in pitch-darkness. Nothing was operational at that point;

all the staff could do was try to get to the plant to check on conditions and report back up the chain to the ERC. Mountains of debris had flowed in and inundated vital equipment. The routes leading to the site were no longer navigable. They had no idea how they would get out, even if they wanted to.

The operators considered their options. The team understood that they were in unknown territory and that there was no way of guaranteeing their safety. The best they could do was proceed with caution and try to stay safe.

Izawa didn't share the fact with his team, but he was extremely frustrated with the situation. As a trained operator, he knew that the best way to control the reactor was to lower the pressure and inject water, and so did the ERC. But he had to wait for input from the restoration teams before he could put a plan into action. Superintendent Yoshida, Inagaki, and their team were trying to get a handle on the situation from the ERC, but they kept encountering mayhem wherever they looked.

ERC CHAOS

When the tsunami came and was much larger than anticipated, operations people in the ERC began shouting, "Unit 1! Unit 2! SBO!" All the emergency diesel generators had shut down and the reactors went into SBO mode. Of course, Inagaki was familiar with SBO, but only in theory—he'd never experienced one and could barely believe it was happening. He heard that all the control panels were going black and silent. He reached out to an extremely experienced and competent maintenance person—a god of maintenance—about restoring electrical power, but found him to be as overwhelmed as he himself was. It was new territory for everyone.

Everyone was shocked by the news that one of the heavy oil tanks had come unmoored and disappeared into the sea. A very large surge tank designed for storing radioactive water during a reactor maintenance shutdown had been seriously damaged and the unthinkable was becoming their reality: the release of radioactive material into the sea.

The windowless ERC became very noisy as they all began discussing the declaration of nuclear disaster they knew they would have to make. Should it be an Article 10, which is a kind of "condition yellow," or an Article 15, which is a "condition red" confirmation of an impending nuclear event? The plant superintendent, Yoshida, was intent on getting as much information as possible to government officials and the media, but he hardly knew what to report in the immediate aftermath of the wave. They spent thirty minutes to an hour working on this decision. At 3:42 p.m., they reported an Article 10 SBO to the government, and at 4:45 p.m., they upgraded it to an Article 15 (radiation release probable) incident.

When an operations general manager came to Inagaki to report the SBO and the loss of instrumentation, it fell to him to set priorities and next steps for his team. The restoration of electricity seemed like the most pressing problem, so he asked the group managers in charge of electrical maintenance to check out all the electrical devices, electrical panels, and emergency diesel generators and think about ways they might restore power.

The operations department reasoned that it was most important to restore the reactor water-level instrumentation, followed by the gauges that monitored reactor and reactor containment pressure. These three things were most important. Then Inagaki asked the instrument and controls (I&C)[2] manager to come up with a way to restore these instruments.

After a short time, the electrical maintenance people approached Inagaki and told him that they wanted to go to the electrical switchyard that housed the high-voltage power supply to the plant for the Unit 1 and 2 reactors. Because it was located some 115 feet above the sea, they assumed it must be safe. Inagaki allowed them to go.

The aftershocks continued over the next several hours, along with tsunami alerts. When some of his workers asked Inagaki if they could go to the electrical switchyard for Units 3 and 4, he hesitated. The switchyard was located only thirty-two feet above the sea, and he knew it would be a dangerous mission. At first, he told them no. He just didn't want to risk their lives. A little later, when they asked him again, he relented. He told them that at least two maintenance people, together with members of operations and radiation protection groups, had to go, and that they should check the sea; if they felt they were in imminent danger, they were to come back immediately.

These brave individuals went off to check the electrical switchgear of Units 3 and 4 and then went into the basement of the turbine building to check the emergency diesel generators, which they found damaged and inundated with seawater. Perhaps they could salvage a 480-volt power center downstream of the switchgear. They reported that possibility, but there was no external power, and the flooding had ruined the emergency diesel generators. There was no way to recover a functioning electrical power system.

"PLEASE DON'T GIVE UP!"

Just when the electric maintenance team had given up any hope of restoring power, a call came in from TEPCO Headquarters. "We are sending power supply trucks via the Self-Defense Forces," they said. "Please don't give up." TEPCO Headquarters went on to ask them to think about how they might use the power supply trucks. They were going to have to find a connection point and establish some procedure for transferring power to the usable 480-volt power supply centers in the plant.

But how would the trucks get through to them? The answer to that question astonished Inagaki. A power supply truck was going to be delivered *by helicopter* and would arrive within fifteen minutes. This possibility gave him a ray of hope. He and his crew waited and waited. They checked with headquarters again. After several garbled communications, they were told, "It will be there in ten minutes." Again, nothing. After several hours, the call came that the truck was just too heavy for the helicopter. It would have to be driven to the plant, through traffic jams and road damage. They had no idea how long that would take. Inagaki struggled to contain his disappointment.

The first power supply truck that arrived was from the Tohoku

Electric Power Company, a neighboring utility. Its driver had no idea how to get to the plant, so Inagaki got on the phone to give him directions. A drive that would have taken ten or fifteen minutes under normal conditions had taken the driver more than an hour. More trucks, at least seventy-two, would arrive from various TEPCO branch offices, as well as from local Self-Defense Forces.

Meanwhile, the I&C maintenance people had figured out how they might recover the most important signals, including that of the reactor water level. They explained that they could revive the water-level instrument if they lined up two batteries and connected them to the DC terminal. There was just one problem—they had no usable batteries.

JUDGMENT UNDER EXTREME PRESSURE

Izawa had always been interested in the experiences of firefighters and emergency workers in extreme crises, such as during America's 9/11. He had read books and seen documentaries on various accidents, including Apollo 13, and had studied official accident reports. He had immersed himself in the accounts of the Three Mile Island accident, the Browns Ferry fire, and the Chernobyl disaster. He had studied how their leaders had responded to the crises.

When he became a shift supervisor, Izawa could not imagine that anything like these events would ever happen on his watch. But, as his career progressed, he came to realize that these things can happen anywhere. Even simulator training might not be enough to ensure proper behavior in such a circumstance—to embed the proper mindset into the leaders in question.

He had told his people, "If you want to be a supervisor, you have to believe that anything can happen." He knew that it took a certain kind of person to train for and handle such a situation, and that those people would be hard to find.

HOPE FOR COOLING THE CORE

The reactors desperately needed water injection to prevent core damage. With reactor pressure high, only the high-pressure injection systems would work. For Unit 1, that meant an isolation condenser (IC); and for Units 2–6, a high-pressure coolant injection system[3] (HPCI) and a reactor core isolation cooling system[4] (RCIC), along with smaller injection systems, control rod drive (CRD), and standby liquid control (SLC). Without electrical power, CRD and SLC would not work. The other low-pressure water injection systems were also without power. The only option at that point was to depressurize the reactor and use a diesel-driven fire pump to provide water through the fire protection system, which would have to be altered to flow into the reactor. But reactor depressurization is difficult to accomplish without the power needed to open the SRVs.

The operators noted that the indicator lights for the diesel-driven fire pump (DDFP), which operated on an independent battery, were on. Could that mean that the DDFP was available for use? Could they align the fire protection system piping and use the DDFP to pump water into the reactor? That would be the first positive development for Izawa, who had started to become disheartened about their course of action.

With the extensive damage in the plant to the fire protection piping systems, this option would be no cakewalk. Further complicating this arrangement was that the systems traverse between the reactor and turbine buildings. The operators discussed the consequences of failure. Then they put their fears aside and began checking to see whether the DDFP would work.

The in-plant operators were responsible for getting water to the reactor core using the emergency core-cooling system piping connected to the fire system piping. If there had been electrical power, it would have been simple to switch between the two systems with the flick of a switch. As it was, they would have to close and open valves, large valves located in a very hazardous environment, one by one, and that would take time. During that crucial time, the core would be without a vital water supply. In addition, there was a reluctance to send the younger operators—who were more vulnerable to radiation—to the field.

The next thing that happened was that the IC on Unit 1 gave out. The IC was a critical safety system designed to help prevent core damage. It would trap water as vapor/steam and return it to the reactor. In other words, the IC continually supplied water to the reactor, while sending residual steam out of the building exhaust system in a big cloud. With the loss of power, it was impossible to verify whether the steam-driven reactor core cooling systems, the IC at Unit 1, and the HPCI/RCIC at Unit 2, were up and running.

Leaders at the ERC assumed that the IC was still functioning as it was a passive system, but it was a safe bet the RCIC, a steam/electrical/mechanical system, was in trouble. Ultimately, all was speculation, and as a result, the ERC had few certainties to share with the control room.

The people in the control room did not have a view of the IC steam exhaust. Crucially, on this issue, there had been a "gap" in the training of operators—they had never seen what it looked like when the emergency IC was in operation. In training, they had come to believe that the IC would release steam in a plume thirty, sixty, ninety feet high. They understood that they would not be able to see the steam from where they were but that they would hear it. On March 11, that is exactly what happened.

Under normal circumstances, steam was supposed to exit near the Unit 1 reactor building, but because he knew he was in a blind spot and the tsunami had washed away a lot of electrical and monitoring equipment, Izawa had asked those at the ERC to inform him if they saw that the steam had stopped flowing. The ERC told Izawa that the steam was billowing out in a "misty fog." *A misty fog? Is that thing working?* wondered Izawa. He knew that it should be forming a great plume.

When the operators heard about this "misty fog," they knew they had a problem. The ERC had reported "steam going out," and presumed it was the reporting that was faulty and not the IC. But as all good leaders do, Izawa paid attention to even the subtlest details: He determined the reports to mean that the IC was *not* working, not because of the operator report, but because he had not been able to *hear* the release of IC steam at all since the tsunami. (Before the tsunami had struck, they could always hear steam.)

Immediately after the plant shutdown, between the earthquake and tsunami, the IC had kicked in and sent steam pouring out of the reactor building. In fact, the operators had worried that it was overcooling the reactor and had shut down the system. In all the ensuing chaos, the ERC personnel missed the fact that the ICs were not in service; therefore, they did not communicate the IC status to senior managers at the site and at corporate ERCs. Preparations were underway to augment core cooling using a diesel-driven fire pump. They based their priority on the incorrect assumption that the ICs were operating and were providing cooling; they were therefore more concerned about providing core cooling for Unit 2. In fact, urgent attention was most needed for Unit 1.

In the ERC, Yoshida was focused on starting water injection into the reactor and could not pay sufficient attention to the implication of the IC stoppage. Without IC operations, core damage had probably begun or would begin soon. Operators were dispatched from the control room to verify IC status locally, but the lack of proper radiation protection equipment and personnel safety concerns caused by insufficient lighting, debris, and ongoing aftershocks prevented them from reaching the ICs. The misunderstanding in the ERC about the IC status continued throughout the evening.

On March 11, around 4:55 p.m., while they were going back and forth about the IC, some of the operators who had gone out to check the facility discovered a route that could be used to get operators into the turbine building. The problem with navigating the landscape was that big aftershocks kept occurring—so big that they had changed the coastline. That sent debris flying and forced the workers to run back to safety.

At 5:12 p.m., Yoshida, still in the ERC, ordered the initiation of alternate water injection from the Make-up Water condensate system or fire engines. At this point, there was only one working fire engine; the other three were either inaccessible or had been destroyed. At 5:19 p.m., a group of operators went to check on the condition of the DDFP. At 5:30 p.m., they determined they could start the pump. Because the fire water valves were not yet aligned to inject water to the reactor, they placed the DDFP on standby. (Operators had to hold a push button down for hours to keep the pumps from automatically restarting.)

DIFFICULT MANEUVERS

At 6:35 p.m., a second operating crew headed out into the reactor building to align the fire protection piping to the Core Spray system piping, enabling the fire protection water to reach the reactor core. Both crews were in total darkness, with only flashlights to guide them. (This situation reminded me of Browns Ferry, where the operators courageously went into a dark and smoke-filled reactor building with a fire hose to put out the fire.) In the reactor building, numerous motor-operated valves on the Core Spray piping had to be manually opened. The handles were large and the valve stem strokes long, making them difficult to maneuver under the best of circumstances. Although the crew knew the fire protection pipes in the Turbine Building went to the reactor, they didn't know whether they had sustained any damage, so they had to check the Turbine Building pipes one after another, foot-by-foot. Some had been damaged by the waves, and they had to decide where to isolate the breaks. It took hours to manually open and close valves that would normally be operated electronically. It took four people to open each valve. After just a few minutes, sweat filled the full-face masks worn by operators. The workers had to choose whether to pull off their masks, facing an intake of radiation, or swallow the sweat. Nevertheless, they persisted, taking turns opening the valves little by little. It is hard to describe how arduous all of this was.

COMMUNICATIONS WITH THE ERC

Three hours after the tsunami knocked out power, and without specific instructions from headquarters, the operators were attempting to inject water into the reactor with a diesel pump designed to fight fires. They had decided it was their only option. Inside the control room, the IC wasn't the only topic of communication with the ERC. Izawa was also discussing firefighting operations, reporting on other conditions on-site, and more. Of course, the ERC wanted to know everything, but the amount of information to pass on was overwhelming.

Izawa assigned an operator the task of communicating with ERC so that he could focus on reactor operations. A lot was resting on Izawa's shoulders, and he was feeling the strain. When someone at ERC wanted to know how much battery power they had left, along with other details, Izawa lost control. "I f****** told you already that we don't have any batteries!" he shouted into the phone. "I told you that we've lost electrical power! What's wrong with you?" But the ERC continued to pose unhelpful questions and communications remained extremely strained and testy.

As night fell, they lost whatever natural lighting they had. Izawa figured that the staff at the ERC must certainly be working on ways to get rid of the debris and, more important, restore electrical power. Right? Wrong. When Izawa tried to communicate with them about his staff's efforts on the diesel fire pump, the ERC staff misunderstood and thought Izawa was trying to align water to the IC. *No, we are trying to get water into the reactor, you idiots,* Izawa thought. Honestly, Izawa couldn't figure out what they were doing. He continued to inform them about progress with the diesel pump. When he told the ERC staff that a valve in the main fire water line was shut because of a leak, again, they thought he meant the main pipe to the IC, and followed up with a nonsensical question. This was not effective communication.

Izawa was thinking about the people on-site and those in the ERC, who he knew were desperately trying to do something. He believed they would do everything they could to assist the control room operators and that they should persevere just a little bit longer. He told everyone as much, making it clear that he wasn't giving them an order at this point, but making a request. Some young operators had wanted to evacuate soon after the SBO, but Izawa had promised them that he would not put them in a situation that jeopardized their lives. He also said that because there were

Restoration Group technicians sent by Yoshida and Inagaki in the plant, working to recover electricity and core injection, the operators had to support them from the control room. Izawa asked them to hang on a little longer, and all agreed to do so.

In time, some operators became too scared to work. Some were compromised mentally, some physically. Eventually, one of them came to Izawa representing the younger operators and asked that they be permitted to evacuate. "Chief, is there any reason for us operators to even be here?" he said. No one responded at first—what could they say?

"You've been brave," someone finally said.

Izawa's response was that the people who'd evacuated the cities were counting on the operators to handle the situation—to save them and their homes from a radioactive plume. Once more, he assured them that he would evacuate them when he believed it was necessary. In the classic Japanese fashion, he bowed and apologized to his operators. This move was less about leadership than a genuine plea for help. He was impressed that the young operator had mustered the courage to speak up, but was also thinking about the local workers, the ERC team, and everyone else on site. Two senior shift supervisors also bowed to the operators.

They didn't ask about evacuation again until the morning of Day 2.

The operators had conducted exercises to rehearse for situations in which damage to the core might be imminent. In theory, they knew what to do. But Izawa intentionally did not openly talk about the reactor status. He did not mention the percentage of core damage, or that the pressure of the containment vessel was continuing to go up beyond what it had been designed to withstand. He deliberately did not share what the data told him—that the reactor had reached a catastrophic point much more quickly than he could have predicted—because he knew they would panic if he did.

At the point that he realized how critical the situation was, his main focus was on deciding when to evacuate. He later reflected: "I was just praying to God. I was crossing my fingers that the containment would survive. Then, due to the efforts of the operators, or maybe Mr. Inagaki and his team's effort, or maybe just timing—I can't be sure—the pressure of the containment vessel went down. I still don't know why that happened, as I never received a report saying the mission had been successful."

THE CHALLENGES MOUNT

The team desperately needed supplies, including radiation measuring instruments (the kind you can keep in your pocket or on your person, as well as the larger ones used at the site), radiation masks, filters, lights, batteries, air cylinders (in the worst-case scenario, they would need them), and so on. Not to mention a working toilet. Remember that there was no running water. There was a working bathroom in the ERC, but that was of no use to those in the control room. They didn't know when transportation people would be available to get more equipment to them. It could be an hour or two, or even longer. The tired crew had no idea what calamity might befall them next.

ON THEIR OWN

Daiichi had been hit hard by the tsunami, and flooding had wiped out their power supply facilities. Izawa thought, *What was Tokyo headquarters thinking when they received requests for help from Daiichi?* Maybe that they didn't want to risk any more lives. The crew at Daiichi had the feeling that they were probably on their own and had to make do with what they had. (Eventually, others from the Kashiwazaki-Kariwa Nuclear Power Plant arrived to help.)

Inside the building, there was a complicated arrangement of pipes and systems designed to supply fire protection water to the plant. The firefighting piping branched all over the building, but with the extensive damage from the tsunami, the pressure was minimal. Even if they could get the reactor pressure low enough that the water could be injected, there still wouldn't be enough water pressure to pump it in.

Years earlier, the Niigata Chuetsu-Oki earthquake had broken this same type of piping at TEPCO's Kashiwazaki-Kariwa Nuclear Power Plant in western Japan. Improvements were being implemented at all their nuclear plants, but they had not been completed at Daiichi. Izawa decided that they needed to set something up outside of the plant to connect for water injection. Studying the diagrams, he'd say, "There should be a valve here." The workers would look for it in vain. At one point, he grew frustrated. "You guys are the ones who don't get it. I absolutely need you to find that valve."

"All right, let's go," they said, splitting into small teams, one for outside the building and one for inside. In the dark of night, the outside team searched through the debris for valves they could use. Inside the turbine and reactor buildings, it was pitch black as well. Staff searching for valves in the basement of the turbine building had to climb over plumbing in the dark, knowing that a false move could result in their deaths. They went on anyway. It would take several hours before the men could connect pipes that would supply water to the reactor. They did everything they could to get the water pressure up high enough to be effective.

Meanwhile, one of the group managers of electrical maintenance told Inagaki that he had found a small generator he thought he could use to restore the lighting in the central control room. "OK, do it immediately," Inagaki told him, and by 8:49 p.m., March 11, they had succeeded. Establishing lighting in the control room went a long way toward reviving the operators' flagging spirits as well as allowing them to work more effectively. It was a symbol of hope for all of them.

Under normal circumstances—including on March 11—there were about twenty-four operators total in the control rooms. On that day, a new crew was due to start at 9:00 p.m. and, much to the surprise and relief of the fatigued and dazed workers, they arrived. How they got there, Izawa never knew; there had been no bus to bring them in. Perhaps they had carpooled. Whatever the case, there they were, and Izawa respected them deeply for their commitment. He'd certainly have understood if they had stayed home.

At that point, they took a little time to strategize and think about what to do. Izawa told some of the younger guys to get everyone some food and water. Surveying the scene, he thought, *they still have it together; they aren't panicking and they're still calm.*

TEPCO is a not a maintenance company, its workers are the operators, thus they didn't keep supplies such as backup batteries and other equipment on hand. The manager of I&C asked Inagaki to help by putting the call out for batteries. The warehouses of TEPCO's affiliate companies were searched—no luck. Then, someone had a brainstorm. Two large batteries were taken from company buses and connected to the reactor instrumentation. Success! The reactor water-level gauges for Unit 1 went live at 9:19 p.m. and for Unit 2 at 9:50.

The instruments showed the reactor's water level to be high, but this reading was incorrect—it was actually dangerously low—and Inagaki suspected as much. "We were quite doubtful about the IC status," Inagaki told me later, "but we hoped that the isolation condenser was still partially working on Unit 1. Because it was the only information we got after getting SBO, we wanted to believe the signal, though we knew the water level must be going down." The signal raised their hopes; they wanted to believe it.

It was much more difficult and took more time to restore the instrumentation monitoring reactor pressure. At around 10:00 p.m., power had been restored to the gauges, and the team's fears were realized: the reactor pressure was high. The operators and ERC workers had no way of depressurizing the reactor automatically to allow the injection of low-pressure water. Opening a safety-relief valve on the reactor vessel was their best bet, but they would need 125 volts, which would require at least ten new batteries. One battery of this type weighed more than 220 pounds—how would they carry these to the control room?

Because depressurization seemed impossible, the ERC leadership strategy was to restore high-pressure water systems. They would need an electric power supply to start either a control rod drive pump or standby liquid control pump to inject high-pressure water into the reactor. Knowing that more power trucks were on the way, they decided to position a portable diesel generator next to the turbine building. Around midnight, they started to lay down the cable under the instruction of the electric maintenance group people. They checked the route to lay down the electrical cable, and a hole was found in the building to route the cable.

At a certain point, the ERC lost communication with those in the field. When a big aftershock hit, a messenger was sent out: Everyone should gather on the second floor of the turbine building for at least an hour. With the onset of more aftershocks and tsunami alerts, however, they all decided to return to the ERC. After the flooding, there was a great deal of standing water around the plant. It was dark outside. The workers had to walk with sticks, probing the ground for open manholes and debris. They made little progress on this dark and moonless night.

At the same time, the I&C people were connecting the batteries and portable generators to recover further reactor instrumentation and controls. They'd been in the central control room for many hours and desperately needed a break to remove their protective masks, eat, drink water. These I&C workers—the operators and the men who laid the cables—were genuine heroes; they were fully aware of the radiation danger to which they were exposing themselves but they did the work anyway, in hope of saving the plant.

Finally, at 8:50 p.m., the alternate injection system was ready to pump water to the reactors using the DDFP. But sometime after midnight on March 12, the DDFP shut down. It had run out of fuel. Someone commenced a resupply effort. Operators tried in vain to restart the DDFP, but the batteries died during these efforts. At 2:56 a.m., they concluded that the DDFP was inoperable. There would be no more water to the reactor cores from the DDFP. This was a depressing moment for all of them.

Around midnight, Yoshida ordered preparations for seawater injection into the reactors. Seawater would cause significant corrosion inside the reactor and would make them unusable ever again. What's more there was a fear that the water injection might make the reactors restart the chain reactor. These fears were dominant among the national government officials and others in Tokyo, but Yoshida was undeterred. He understood that, with limited freshwater supplies, seawater was the only option to keep the reactor cores cool.

CHAPTER 5

NEXT CHALLENGE: VENTING

Then I heard the voice of the Lord saying,
"Whom shall I send? And who will go for us?"
And I said, "Here am I. Send me!"
—ISAIAH 6:8

A bout midnight on the 11th, the I&C people recovered the Primary Containment pressure instrumentation, and it showed very high levels of pressure—much higher than it was designed to withstand. Steam pressure was still building up within the reactor containment vessel, and venting the steam was the only solution. It was then that the leader of the operations team told Inagaki to commence venting immediately—but it was still unclear how.

It was the opinion of the leader of the Operations Department that they'd have to open the valves manually. Accordingly, Inagaki asked the reactor maintenance group manager to review the drawings they had of the system. It seemed that they could manually open the motor-operated valves (MOVs), but the air-operated vent (AOV) valves had no handles on them. The operations people suggested they attempt to use a lever of some kind to open them, but they knew it wouldn't be easy.

The maintenance team began making calls to their valve manufacturers, even the plant manufacturers, for advice and perhaps tools they could use to open the AOVs. Naturally, it was hard to reach anyone, and when they did, the news they got wasn't encouraging. Everyone they spoke with agreed that it would hard to move the valves with the tools they had available.

UNTHINKABLE ORDERS: OPERATORS PREPARE FOR VENTING

Venting in an SBO is a terrible last resort. *I'm going to spread radio-active materials around the area,* thought Izawa. *My family lives here, my friends!* Giving this order seemed unthinkable—but he had to give it. Just after midnight on March 12, Yoshida ordered advance preparations for containment venting.

There was significant discussion in the prime minister's office about this drastic move. Prime Minister Naoto Kan was frustrated that the head of Japan's nuclear regulator, the Nuclear and Industrial Safety Agency (NISA), was an economics graduate, not a nuclear expert. Some advised Kan that there would be no explosions. Kan ordered the venting at around 1:00 a.m. His government was letting him down, and he felt that a trip to the site was necessary. There was much debate about this trip; nevertheless, Kan went to Daiichi. He would become extremely frustrated at the slowness of the venting process.

Thinking back on it, the situation emphasizes to me the value of the resident inspector program at the U.S. Nuclear Regulatory Commission. Resident inspectors can provide direct and independent situation assessments to the head of the government. NISA did not have a resident inspector program, so Kan did not have this advantage.

At Daiichi, it was very tough to select the operators for the task of venting, as it was almost an entirely manual operation—*and a kamikaze mission,* in Izawa's view. He volunteered himself for the job first, feeling that he couldn't send his staff into harm's way yet again. He asked if anyone would come with him. After a moment of silence, young operators who had not even been on his list for the operation volunteered one after another. When I heard this story, it reminded me of the Bible verse Isaiah 6:8: "Then I heard the voice of the Lord saying, 'Whom shall I send? And who will go for us?' And I said, 'Here am I. Send me!'"

Izawa wept at this gesture of selflessness. He had been fully prepared to demonstrate his commitment to the task at hand, but his supervisors intervened. They prevailed upon him not to go—in fact, they commanded it. He needed to stay behind and give instructions, they insisted.

One of the supervisors took Izawa's place. Izawa prayed that he would come back safely. During the early stages of the event, the senior supervisors had not always been in the control room, but Izawa had. He knew he was best suited to continue overseeing the operation and so did his supervisors, so that was that. It would prove to be one of their wisest calls. Izawa chose his words carefully as he told the operators how to prepare for venting. Even at this point, nobody was panicking or breaking down. They were prepared to follow orders. There was no talk at that point of anyone evacuating—they were all in it to the end, including, of course, Izawa himself.

Izawa remembered that, as a young operator, he'd had a shift supervisor who had told him that the person who is the final decision-maker must stay. Even without that instruction, his instincts would have told him the same. At that moment, on that day, although he had his senior supervisor next to him, he was the supervisor—the leader of last resort. Izawa had read many books about the Bushido code of the samurai and had internalized their sense of duty; although, if he'd been asked why he was there, he couldn't have explained it in words.

Those tasked with venting the containment did not know what they'd find inside the plant. They didn't know what the radiation level would be. All they knew, or hoped, was that if they came back safely, everything might return to normal.

Izawa did not want to think about the peril of the situation, but he couldn't completely deny the possibility that his men might die on this difficult mission. They had to put on pounds of protective gear and carry a breathing apparatus that they knew would last less than twenty minutes. They had to be thinking they might never see their colleagues—or their loved ones—again. When he looked back on the events of that day, Izawa marveled at the fact that no one, not even the younger men, panicked. Each of them mastered his fear to do the impossible. "If even one of us had panicked, everyone would have started to panic," he shared.

The men were divided into three teams. If something were to happen to one team, the next would step in to relieve them. As Izawa dispatched the first team to the reactor site, he pondered the consequences should they fail. It would be his failure, he believed. He was relying on this system of backup to get the job done. In retrospect, Izawa realized that it would have been a mistake for him to be the first person at the vent—but he couldn't help feeling sorry for the older operators who'd followed his lead and put themselves in the gravest danger. They awaited the order to vent.

INJECTING THE CORE

Inagaki couldn't remember how many TEPCO firemen were at the scene with water—probably around twenty—but he emphasized that they played a critical role in the aftermath of the disaster when they helped inject water into the core. At around 2:00 a.m. on March 12, they started a search for a connection the firetrucks could use to pump water into the alternate injection lines. The search for that connection was complicated and greatly hampered by the tsunami debris all around the buildings.

Operators had never used firetrucks for this purpose. A kind of connection coupling was installed in the turbine building wall to enable water injection from the firetrucks by way of the fire protection and Core Spray systems. This had previously been accomplished by two operators. Mr. Yoshida, who had once been the general manager of the Nuclear Asset Management Department, knew about this protocol. Inagaki had been the one to oversee its installation. The connection was put in place in case of fire, not for the specific purpose of accident management, but Inagaki sensed it was their answer and knew that Yoshida would understand the process. It had been installed just four or five months before March 11, so very few people knew where it was. (This reminds me of the Browns Ferry connection to the key valves for core cooling. It was generally an unknown connection point that saved that reactor.) When the fire brigade couldn't find it, they asked one of the architectural engineers to guide them.

Fortunately—or unfortunately, Inagaki didn't know for sure—at about 2:00 a.m., when I&C maintenance members recovered the reactor pressure instrumentation, the ERC learned that the pressure was almost as low as in the containment vessel. This could mean that the reactor vessel bottom head had been penetrated, or some boundary piping had broken. Whatever the case, good or bad, they had to keep moving forward.

At 3:30 a.m., they found the water injection point. With the connection to the firetrucks secure, at 4:00 a.m., they used a firetruck to begin pumping freshwater into the reactor. It would be a long process, and they had only a limited supply of freshwater. This freshwater supply would not last long and eventually they would have to revert to using seawater. Inagaki understood that seawater would ruin the reactor, but he resigned himself to the fact that they'd eventually have to use it, just as Yoshida had resigned himself to the same thing. Soon, a high radiation signal prompted them to suspend this effort. At 5:46 a.m., the injection was restarted.

THE "GO" ORDER: OPERATORS ATTEMPT TO VENT

Unaware that Inagaki and his team were already working on the venting process, at around 8:03 a.m. on March 12, Yoshida gave the "go" order to his operators in the control room to start the venting process at 9:00 a.m.. The first team managed to open one of the valves on the second floor of the reactor building. When they returned to the control room and reported their success, there was elation all around—they were feeling that success was in their reach as they dispatched a second team. Those operators failed in their effort to reach the second valve, located in the Suppression Pool room, because of exceedingly high radiation levels. At that point, Izawa understood that he'd have to come up with another method for opening the valves. As the central control room staff continued working, radiation levels continued rising in the control room, and the operators knew it. They did what they could to limit their exposure without abandoning their responsibilities.

ONE MORE HEROIC TRY

It was the afternoon of March 12, and the control room operators had no intention of giving up on their efforts to vent. They begged the ERC to do something about their electrical power. "Please think of some way you can open the vent valves from the outside to let the steam out," they pleaded.

Their best bet would be to figure out how to open the air-operated vents remotely, and they continued working on this problem. Inagaki exhorted the I&C group manager to try to use electrical power to make this happen.

The ERC people hoped that the instrumentation air system had some residual pressure in it they could use to open the air-operated vents. Operators tried three times to open the valves from the main control but failed each time. The air pressure just wasn't sufficient to do the job.

Inagaki persisted. This time, he asked the turbine people to try to find a way to connect a diesel-driven air compressor to the instrument air system and send in pressurized air. They quickly found the connection location but had no idea what kind of adaptor they needed. The team leader told Inagaki that he needed to go out to the site and search for an adapter. Inagaki agreed, but insisted that parts supply people accompany him, along with others from his ERC team. They quickly determined what kind of adaptor they needed, then went to the warehouse and found it, along with the best air compressor for the task.

Next, they'd need a truck with a small crane. None of the operators knew how to drive such a truck—and they had no ability to communicate with the ERC. Given that, they had to proceed slowly and cautiously. They felt daunted at every turn, but no one even spoke of giving up. They knew they'd have to answer to Inagaki if they did. After a conversation between Izawa and the ERC, the operators, unwilling to give up, decided to attempt venting for the third time. Again, Izawa faced the decision of who to send.

Up to this point, operator age was the basis for the selection. Everyone who had gone out was in his fifties. Now, some younger guys from the other units were volunteering to help. They were vital and fit. Izawa decided that a fresh, young team might be better.

Izawa tried to persuade the older men to step aside, but they simply refused. One of the senior guys refused to take off his fire protection suit and give it to a younger operator. "Do you know what it's like down there?" he asked them, but he understood that they were prepared to die. In the end, it was the youngest men who headed out; Izawa was certain that they were vital to the success of the third venting attempt and believed they were ready for whatever might happen. He learned later that they'd gone in without even taking air radiation detectors, though they did have personal exposure monitors.

At 2:00 p.m., the turbine workers succeeded in connecting the air compressor and sending air into the installed instrument air system. Before long, from the ERC, they could see a steam plume rise above the stack tower on the television monitors. It seemed that their efforts to vent the containment had paid off, but they had no way to let the operators know this.

Just as the two young operators went off to attempt to open the vent valve, Izawa got a call from the ERC that there was "white smoke" coming from the plant ventilation. Not knowing that the turbine workers had succeeded in starting the vent, the ERC asked if the situation had changed. *Something is going on in there,* Izawa realized. The team he'd just sent into the plant was going straight toward something dangerous. For the first time, he outwardly panicked. "Stop them! Stop those two!" he shouted. Just before they entered the reactor building, another operator stopped them and turned them back to the control room.

CHAPTER 6

DESPERATION

---

*"You are killing the people! You are killing the people!" he
shouted. Inagaki says he can still hear Yoshida's words
ringing in his ears."*

A
t 2:00 p.m. on March 12, after great struggles, the power supply
people succeeded in connecting an electrical power supply
truck. The freshwater supply was dwindling fast, and at 2:54
p.m., Yoshida ordered the start of injecting seawater. Workers began looking for a seawater supply point and laying hoses to begin
that injection. At 3:30, they were just about to start a standby liquid
injection high-pressure water pump. At 3:36 p.m.—exactly twenty-
four hours after the second tsunami crashed over the seawall—as
the central control room operators stood by in horror, the Unit 1
reactor building exploded.

The staff in the ERC thought at first it was just another aftershock—but aftershocks usually involve a large jolt followed by
smaller ones; this jolt was singular. "What happened?" they asked
one another. Someone posited that the main generator must have
blown up, knowing that it contained hydrogen. Another suggested
that it might have been the hydrogen cylinders. But when they
glanced at the televisions in the room, they saw a charred skeleton
where the reactor building had been.

"What is that?" Inagaki said aloud, unwilling to believe his eyes.
A moment later, the news channel replayed the moment of the

explosion. There was no denying the horrible reality of what had happened. The bustling room went silent.

In Tokyo, Kan and his leaders watched the explosion as well. The academics who had told Kan there would be nothing of the kind, stood mute. Kan wanted to know if this was a "Chernobyl-type" explosion.

In the pitch-black control room, in full face masks, the men were essentially blind. In addition to the reactor building, the explosion had destroyed the fire hoses that the workers had just laid and injured several of the heroic electricians still laying cables. Workers who had fallen back after the explosion returned to the site at 5:20 p.m. to find even more debris and even higher radiation levels. At 7:04 p.m., they succeeded in starting seawater injection into the reactor.

That evening, there would be disagreement among the ERC, TEPCO leadership, and the prime minister's office about the use of seawater. This is a well-known aspect of the story, but I'll say for the record that Yoshida's decision to continue seawater injection—even though he had to lie about it—was the right one, notwithstanding the Japanese cultural dilemma regarding following orders. At that moment, Yoshida was an operator—not a *Japanese* operator. He knew that seawater injection was the right course to take, and he assumed responsibility.

When I asked him about covertly defying the order, he told me that he knew if he had openly flouted it, his bosses would have replaced him with someone who would follow their protocol—and that would have threatened all of Japan and beyond.

IT'S TIME TO EVACUATE

The explosion of Unit 1 caused the control room ceiling to collapse partially, and it looked like it might come down altogether. People and objects collapsed to the floor. What fluorescent lights they had left stayed on for a few moments, then flickered out. Once again, the men were in total darkness and completely disoriented. Someone got on the phone to the ERC to ask what had happened, but they were still trying to figure it out themselves. Even before the hydrogen blast had ceased to echo through the chamber, Izawa had decided that it was time to leave the control room. For the moment, they didn't know what was happening. Ironically, the world was watching the events unfold on television, while the operators themselves could not see a thing.

Izawa knew for certain that it was time to leave the central control room and ordered an evacuation, but they all knew that some people would have to stay behind. They discussed the matter and chose a team and a substitute team, but again, the senior supervisor from the Units 1 and 2 central control room intervened, telling Izawa: "If anyone knows what's going on here, it's you. Will you please go up first and talk to the ERC directly?" Izawa understood that this wasn't a request, it was an order from a senior supervisor. He, in turn, ordered some staff, including all the younger workers, to head toward the ERC with him.

Someone suggested taking a picture of those who would remain in the central control room. In Japanese culture, it is considered bad luck to take a picture in such a situation—and there was no light to illuminate such an image—but he snapped it anyway.

Later, Izawa would remember hearing someone ask the photographer, "Why did you do that?"

The man replied, "After Unit 1 exploded, I figured Unit 2 would be next. Maybe it would even reach Unit 3. Who knew how long the control room would remain intact?" He'd wanted to capture an image of those brave men in case they didn't make it out (Figure 4). "If they all died and someone found the camera, at least there would be a record of those who were there," he said. That photo exists today, and in it, you can see a supervisor's chair. It is Izawa's chair, but he is not in it. The operator who was didn't even bother to look at the camera.

Figure 4. Inside the control room

Izawa and the others he'd chosen made their way to the earth-quake-proof ERC, which was wide and spacious compared to the cramped quarters of the control room. He entered through the one available doorway and made his way up the stairs to the second floor, only to find people sleeping under desks at which others were sitting, paying no attention to those sleeping under them.

There was a whiteboard with notices for each Unit. When Izawa saw it, he said, "Hey, they haven't written everything down. They didn't write down that Unit 1 is having major reactor trouble and can no longer be restored easily. It's time to start discussing the most desperate measures!"

The people there, ready to relieve those in the field, were duly alarmed. "Why should we have to go to the site if it's so dangerous? Why are we being dispatched there?" they cried. These were not operators. They were other personnel, completely untrained and incapable of doing the assigned work. Some, Izawa suspected, wouldn't be able to handle the situation emotionally in any case. Their fear would get the better of them. Izawa sensed that the ERC had no system for relieving personnel; they were simply grabbing anyone they thought they could use. He found Superintendent Yoshida, but couldn't speak with him as he was constantly on the phone. *Who is he talking to?* Izawa wondered.

Yoshida stood out as a large man. Izawa found him angry, aggressive. Later, many stories were told about Yoshida's actions during the disaster, including a moment when he stood up, turned his back to the camera, and dropped his pants. He pretended that he was re-tucking his shirt, but everyone knew what he'd meant by showing his backside to his superiors at headquarters. Often after speaking with someone from Tokyo, he would slam the phone down, look at it, and shout, "Idiot!"

Information on each Unit began to come in, estimations of how bad things would get and where their priorities must lie. Izawa was never sure how much of this information made it to each department.

YOSHIDA TAKES CHARGE

Yoshida shouted, "We cannot repeat this with Units 2 and 3! Everyone must report to me precisely what they are doing and what they are going to do, and by when!" At this moment, he became a kind of god. He, alone, would oversee the operation on the ERC side from this point on. Per his orders, a group designated as the Restoration Group gathered around a whiteboard and began planning the restoration of seawater injection into the reactors. The commander of the original fire brigade—Inagaki's deputy—had been seriously injured in the explosion, and the power supply trucks and equipment had been damaged as well.

The restoration effort was focused on Unit 3. The instrumentation was still working there because the batteries powering it were located somewhat higher than those for Units 1 and 2. But, because water had flooded the battery charger, they knew that before long they would lose all the instruments. At 8:36 p.m. on March 12, hours after the explosion in Unit 1, they lost all signals for reactor water levels in Unit 3, but by then the team had come up with a way to restore the instruments. Inagaki asked the I&C people to prepare to vent the Primary Containment of Unit 3 immediately. He directed them to install the batteries that would recover instrumentation for Unit 3, as had been done successfully for Unit 1.

Tokyo headquarters had sent new batteries, though their voltage was low. To get 24 volts, they'd need twelve batteries weighing about sixty-five pounds each. These would have to be carried by hand, one at a time and in total darkness, up to the control room on the second floor. The men reported later that they'd grown exhausted after carrying just one battery, but they stayed at it and completed the task. By midnight, a day and a half after the first earthquake, everyone was on the brink of collapse. Some had faded out of consciousness and were sleeping in their chairs.

They were jolted awake at 2:42 a.m. by the news that HPCI, the main water injection system, had shut down around midnight. The shutdown of HPCI meant that the core had been without water for several hours. That bad news instantly galvanized the tired crew back into action. Inagaki asked the electric team to accelerate their efforts to recover power. After that, he led a very long discussion about ongoing strategy. Eventually they would find more than a thousand 12-volt batteries at a location 35 miles from the site. Unfortunately, no one was brave enough to drive them to Daiichi.

TRYING TO SAVE UNITS 2 AND 3

Initially, they decided that the next step would be to restore any of the high-pressure pumps for Unit 3. But as soon as Yoshida realized that the actual trip time of HPCI had been hours earlier, and that the core was in considerably more danger than they'd assumed, he and Inagaki revised their strategy. They had to open the safety-relief valves as quickly as possible, knowing, of course, how arduous the task would be.

After speaking with the I&C people, they determined that, although car batteries were not an ideal source of power, using them was their only shot. Leaders of the Restoration Group, including Inagaki, began asking those in the ERC to produce batteries from their personal and company cars. Some declined to donate to cooperate, figuring they might need their cars to evacuate, but many jumped into action. Ten batteries were amassed for Unit 3 and ten for Unit 2. The I&C people were then told to begin preparing for containment ventilation of Unit 3 first, followed by Unit 2.

It was no easier this time than it had been the last. Thinking that the air cylinder might still be pressurized, the I&C people energized the solenoid valve of the air-operated vent, but could not open the air-operated vents. When they checked the pressure, they found that the cylinder was almost empty.

"Change it," ordered Inagaki, and they set about doing so.

Finally, they managed to open the air- and motor-operated valves of Unit 3, but their relief was short-lived. Moments later, the connection of the cylinder to the air piping sprang a leak, causing the venting to lose air pressure fast. Inevitably, the vent valve shut again, and the Primary Containment pressure went back up.

"Change the cylinder again," said Inagaki. They'd brought another one, but when they attempted to attach it, they realized it was the wrong size and wouldn't work without an adaptor. Quickly, someone went back to the warehouse and found one. They prepared to go back into the reactor building, but by this time they had to take the precaution of putting on cumbersome safety gear because the radiation level was rising quickly. Many prayed that the protection would be enough.

Three or four teams of I&C maintenance people and operators headed out to Unit 3 to open the vent valves. They all fully understood it was their last chance to save the reactor. The first team succeeded; the second team had no luck; and the third team completed its task but under very perilous conditions. They reported back that the connection was perfect, with no leaks. Inagaki heard someone say aloud what they all knew: "But even so, it's quite unreliable."

The replacement compressor was much smaller than the original compressor, and unfortunately, the connection point was very far from any electrical power source. The manager of the turbine crew told Inagaki it would take a very long time to get the air where it was needed. After many hours, the pressure inside the Unit 3 containment finally started going down, but it was unclear if the compressor was even working—much less working well enough. Ultimately, the air in the cylinder ran out, and the venting process was a failure.

The Unit 3 reactor operators prepared for water injection via fire hoses, as had been accomplished at Unit 1. Using the ten batteries they had, they succeeded in opening the safety-relief valve, but, once again, the valve was unreliable. The water injection was continuously interrupted and the reactor pressure kept fluctuating, which meant that, basically, everything that should be under control wasn't. Because the high-pressure injection systems in Unit 3 seemed to be a little more stable than the Unit 1 isolation condenser system, Yoshida prayed that Unit 3 wouldn't suffer the same fate. He was desperately worried about the workers in its shadow, constantly asking them for status reports and telling them to return to the ERC if their instincts told them they were in imminent danger.

On March 14, at 11:01 a.m., the Unit 3 reactor building unexpectedly exploded. Yoshida and Inagaki were at once horrified and crestfallen. Fifty of their bravest men were on the scene—would any of them make it out alive? *I am the one who should be martyred,* thought Inagaki. At that moment, the video system captured him holding his head in his hands.

Yoshida called frantically to his men to return to the ERC immediately and—by some miracle—they all made it back. They were pale and shaken, some were bleeding, but they'd all survived.

Yoshida reported to TEPCO Headquarters: "'Head office, head office! Big trouble. Big trouble." How much radiation was released was unknowable. Yoshida addressed his people: "We're in deep trouble," he said, "but let's calm down. Let's all take a big breath. Inhale, exhale."

Because the Unit 3 explosion looked different from that of Unit 1, there was a lot of confusion about whether the explosion was a nuclear one, as at Chernobyl, rather than another hydrogen blast. What no one realized at the time was that Unit 3 contained "mixed-oxide" fuel, meaning that the fuel contained plutonium, thus accounting for the difference. At this point, Yoshida had everyone scrawl their name on a whiteboard so that history would record the names of "the Fukushima 50."

The Unit 2 reactor continued to present a series of its own nightmares. After the explosion in Unit 3—and possibly because of it—at 1:25 p.m. on March 14, the Unit 2 RCIC failed, and the water level started dropping. The operators relayed this to the ERC, which was still reeling from the explosion of Unit 3. Summoning up all his strength, Yoshida begged the uninjured workers, "Please go again to the field." He could barely believe he was asking this of them, but he didn't know what else to do.

The containment air-operated vent valve for Unit 2 had been open before the blast, and that had felt like a real coup. Now, the vent was closed, and Yoshida wanted to know why. Inagaki told the men to check the cylinder pressure and the circuit they'd put in place to energize the valve. They confirmed that they still had sufficient air pressure. When the I&C technicians checked the circuit, they saw that the valve couldn't be opened further. They feared that even if they managed to open the safety-relief vents (SRVs), they still wouldn't be able to depressurize. They'd need to open the ventilation air-operated vent or there simply wouldn't be enough pressure relief.

At this point, a call came in from a very high place: an eminent professor at Tokyo University who served as the chairperson of the Nuclear Safety Commission. "The valves must be opened using force," he said. Yoshida was dead set against this. He asked his men to try to open the SRV—but they couldn't budge it. "Explain why!" shouted Yoshida at Inagaki, but of course he had no answer. "This was the hardest moment for me," Inagaki recalls. "It felt like I had a stone in my stomach. I looked at the I&C manager and found him sitting like a Buddha, not knowing what to do at this point. Every few minutes, we got a call from the general manager of the Nuclear Asset Management Department, asking for an update. He asked me things like, 'Is this system voltage-oriented or current-oriented?' I'm a mechanical engineer; I couldn't even answer!"

The ERC asked the I&C people to add one more battery to the circuit and then report back with the result. After about an hour and a half, they reported that they'd succeeded in energizing the SRV and the pressure started going down. When he heard this, Inagaki fell back into his chair in relief. They'd made some progress—but the pressure had gotten very high and was dropping very slowly. They tried to open one more SRV and succeeded. Sufficiently low reactor pressure was achieved to inject water from the firetruck, but where was the water level? The other leader of the Restoration Group in the ERC provided an answer. It seemed that the firetruck had run out of fuel. Preoccupied with the rising radiation levels in the field, no one had checked the fuel level in the firetruck. Yoshida was livid. "You are killing the people! You are killing the people!" he shouted. Inagaki says he can still hear Yoshida's words ringing in his ears.

HUMAN MELTDOWN AND DESPERATE MEASURES

Yoshida had seemingly reached his emotional limit. At one point he said, "We have neither water nor ideas." He began wandering around the ERC as if in a trance. Then he slumped in a daze, looking as if he might pass out. This alarmed everyone. Had they lost their leader? Some ERC members walked him to a private room and locked him inside. He would be there for the next two or three hours.

Try as they might, the field teams could not gain control of the SRVs. They needed to open two containment air-operated vents in Unit 2 to recharge nitrogen to the accumulators, including one just next to the nitrogen cylinder in the annex of the reactor building. Finally, an experienced operator did manage to open it and reported this news to Inagaki, who was elated. He immediately dispatched the reactor group managers to try to open another SRV. When the managers opened the door to the reactor building, a steamy fog enveloped them, along with unimaginably high radiation levels. They left immediately.

On March 14 at 11:35 p.m., the operations people concluded that they should try to open the Primary Containment valves above the water, although they knew that this meant there would be no possibility of removing the radiation. "Please allow them to be opened," Inagaki implored Yoshida. The safety people in the plant agreed, but those at Tokyo headquarters were not certain. There was a typically long discussion of the pros and cons, and the conclusion was that they should open the valves. Again, the operation failed.

It didn't seem possible that things could get any worse. When a manager at Tokyo headquarters asked the ERC team to report on the status of the plant every minute, Inagaki almost lost it. Everyone's nerves were frayed, and bickering broke out. Finally, in the early hours of the morning, Tokyo headquarters directed them to watch their teleconferencing system: The prime minister would be arriving at headquarters to integrate command between the government and TEPCO.

No one in the ERC could imagine what kind of speech the prime minister was about to make. They hoped for words of encouragement, but the reality was quite the opposite. When he took the microphone, the prime minister commenced shouting, blaming, and criticizing everyone involved. They later told me that this was not the kind of leadership they needed at this point; it was not real leadership at all.

After the prime minister's rant, the despair in the room was palpable. Yoshida, who had returned from his isolation, was conspicuously silent. Another jolt—albeit milder than the earlier ones—followed, causing the low Suppression Pool pressure in Unit 2 to plummet suddenly toward zero. The moment that happened—indicating a failure of the containment and possible major release of radioactivity—Izawa was notified.

Yoshida immediately gave the order to evacuate the ERC. Inagaki told me that it might seem reasonable to lose hope and give up under the circumstances, but he never entirely did. He was, however, relieved that so many people were no longer going to be in such dangerous conditions.

Yoshida told everyone but the general managers that they were free to leave, but Izawa had no memory of his ever saying, "Head to Daini." In Izawa's recollection, it wasn't that specific. *Was it a permanent evacuation or just a temporary withdrawal?* Whatever Yoshida had meant by his order, it was clear that the panicked staff was all too ready to get as far away as possible.

Predictably, the evacuation was anything but orderly. Hundreds of workers headed for the gate at the same time, clogging the stairs and shoving others to move faster. Some took masks or other protective gear, and scuffles broke out over this equipment. Some of those staying behind secreted these things so that they would have them later.

Inagaki summarized the situation for me this way: There'd been a jolt; Unit 2 was in trouble; Suppression Pool pressure had suddenly dropped; the Primary Containment itself was in jeopardy. How could Yoshida expect them to stay when Primary Containment could fail at any moment? And if it did, a significant amount of radioactive materials would be released into the environment.

After what was termed a "temporary evacuation," the ERC grew very quiet. Radiation in the building measured one millisievert per hour, a nonlethal level. "Do what you want," Yoshida told those remaining. He instructed them to take some time to decompress, or perhaps pray. Some managers refused to give up and attempted to keep working. Some volunteered to check on the fire engines, air compressors, or other equipment.

Inagaki, along with another manager, went to Units 2 and 3 to refuel the fire engine. Since this was their first time going out into the chaos after the earthquake, they couldn't locate the gas or hand pump. They were stunned to see black smoke billowing from Unit 4, which had not even been operating at the time of the disaster. *What could be causing that?* Inagaki wondered. From his vantage point between Units 2 and 3, he realized the source of the smoke was a fire on the fourth floor of the reactor building. *Another explosion?*

The two managers returned to the ERC to report to Yoshida and call the fire station for assistance. Of course, no one was there.

Later that day, March 15, the Unit 4 reactor building exploded much as Units 1 and 3 had, but it wasn't immediately clear to those on-site what had happened. They first ascribed the jolt to a problem in Unit 2. Knowing that Unit 4 had been defueled, Yoshida couldn't imagine anything serious happening there. It would take confirmation from off-site to convince him that indeed Unit 4 was lost, presumably because of some connection to the other reactors.

Some of those in the electric maintenance group had now witnessed three explosions, and they were traumatized. After the explosion of Unit 1, they had asked their group manager if they were safe. The manager had asked Yoshida, who had asked Inagaki. Inagaki couldn't possibly answer the question with any certainty, so

Yoshida had to make a judgment call. Both Yoshida and Inagaki had tried hard to make informed decisions, but as the situation had deteriorated, they'd lost confidence in their judgment. They'd hesitated. Sometimes, they'd made the wrong call. One could say their leadership faltered or lapsed, but that was entirely understandable.

To this day, I cannot imagine a more challenging set of circumstances. Both men would spend years thinking about what they might have done differently, but in the moment, they'd been the best leaders they knew how to be.

THE FUKUSHIMA 50

Inagaki would remain in the ERC continuously for more than a month before returning home. Izawa lingered inside ERC, looking at the faces around him. To the young people who said, "I'm going to stay," he said, "No. Go join the evacuation." In what seemed like minutes, there were only seventy people left at Daiichi. They became known as the "Fukushima 50" because, at any given time, about fifty people worked while the others rested.

"It's just us!" Yoshida shouted. To rally their own spirits, they smiled and joked about being "left behind"—and then everyone grew quiet. Surely, they were thinking about their lives on the outside and wondering when or if they'd make it back safely to their loved ones. Yoshida stunned everyone when he looked at all of them and said, "Let's eat, because we can't really do anything else right now." They agreed and each went to get whatever food they could find. They sat around eating emergency rations, commenting on the food's freshness (or lack of it), and toasting one another with "*Kampai*" (the Japanese equivalent of "cheers") before they drank their water.

REFLECTIONS

As I said earlier, in 2013, I conducted interviews with the key leaders at Fukushima. Even several years after the event, they told their stories with deep emotion. Some operators cried as they talked. Some shared with me the messages they and their operators had sent to their families before power was lost.

Izawa had sent a text message to his three children but didn't remember the exact contents. He said it was something like: "Your dad's staying here until the end. Take care of your mom." Just as Izawa pushed the "send" button, Yoshida (who had seen him write the message) said: "This is great. Now, everything is taken care of."

One operator texted: "I thought it was an aftershock or something, when [redacted] said 'get back!' and before my eyes, I saw a kind of explosion at the turbine building; it was the wave crashing and then water started pouring in. That's when I understood that it was the tsunami. [Redacted] and I escaped and ran to the north. I thought I was going to die then." He later told Izawa that since he hadn't heard back from his family, he thought they might be dead.

Another email recorded at the moment they lost electrical power, and things seemed completely hopeless: "Naturally, everyone understands what will happen if these circumstances continue, but nobody says it aloud. We can't contact anyone either, although I did reach my father in Niigata. I asked him to pray for me if something happens."

Another operator sent a message relaying the futility of it all: "Operators are unable to do anything at all. It's unlike anything we drilled for. Doubts had been raised during drills though: What happens if a station blackout can't be corrected within eight hours? What if a tsunami comes or the plant sustains heavy damage in an earthquake?"

Another text: "Units 3 and 4 were hit by tsunami and aftershocks. Turbine building truck door flew apart. Water flooded the buildings, though we didn't grasp right away that it was a tsunami. We ran north to get out of the truck bay. It's OK now."

One operator told his family that if they stayed, the radiation in the control room would kill them. Another reflected, "When SBO hit, despite this, no fear was upon us, no complaining, mission."

Another text stated: "It felt like I wasn't able to use my training. I felt helpless. It felt like both hands tied with no brakes in the car. SBO was the training."

When one operator couldn't contact his wife, he sent a message: "Want my family to take care. I put my contaminated jewelry back on so that you could identify my body."

Another operator expressed a wish for just one working pump. "We're like the imperial navy soldiers isolated from headquarters on a Pacific island."

Another man wrote: "There are so many value conflicts . . . the prime minister's office wants info, the site is busy responding. Prime minister doesn't understand how hard it is, what we're doing. He made us wonder, what is the value of our effort and sacrifice?"

And yet another person explained that even the most routine tasks had become matters of life or death. "When Yoshida asked people to go out there and said it was safe, they screamed at him, 'You're a liar, you're lying to us!'"

It would be two weeks before Izawa left Daiichi. His wife picked him up, and he got to take a bath for the first time in two weeks. As he climbed into the tub, he started shaking uncontrollably. This embarrassed him, but he kept repeating over and over, "You're alive, you're alive." The man had just been through horrors that infinitesimally few will ever know, and he believed that God had helped him through it. He put his head in his hands and gradually began to calm down. Thinking back on this little breakdown, he said: "It was because I'm just a normal person. Once I no longer had to do anything to avert a crisis, I turned back into that regular person."

In my opinion, Izawa is no "regular person." He is a bona fide hero.

In 2013, as I sat with Izawa talking about his experiences, a very experienced TEPCO manager and colleague and good friend of his, Akira Kawano, joined in the conversation. I wanted to know more about Izawa's character, and a lively discussion followed. It appears below, verbatim:

CASTO: Can you share what experiences prepared you for leadership?

IZAWA: It may be a Japanese way of thinking, but in the case of TEPCO operators, it is not unusual to work as an operator throughout one's career. In my case, though, there was a period where I was away from operating. . . . [P]eople working in the central control room are like my family. If you live in that environment, even if you are not aware of it, you are inheriting the mindset from your supervisors. I thank my parents for giving me this attribute.

KAWANO: Of the three nuclear plant sites TEPCO has, Fukushima Daiichi is the oldest, so it is the oldest technology and rather difficult to operate. Experienced operators like Izawa have operated it for a long time, and it has been very challenging for them. In a sense, they must be creative. More knowledge is required to operate the Fukushima Daiichi. In a good sense, it elevates the skills of its operators, but it may make them more stubborn, too. They have more affection for the plant; they take care of the plant as if they were taking care of high-maintenance, bad kids. As people, they are not necessarily easy to get along with, but as you get to know them better, you know that they are good people.

[Izawa and I] used to play tennis together. He is focused. He has a very strong sense of responsibility. On top of having worked at this difficult facility and having dealt with complicated human relationships, all [his attributes] combined to enable him to handle that tough situation.

IZAWA: Operators from the Fukushima Daini and the Kashiwazaki-Kariwa were reluctant to come to Fukushima Daiichi. There is a lot that you must operate manually. Even before the earthquake, we knew that Unit 1 would be decommissioned at some point in the future. I used to say to young operators that I was going to put the last control rod in the Unit 1 to shut it down. That came true, but in a very different situation. I said I would be the last person to see Unit 1 go, and that also came true.

KAWANO: I was far off from his fighting spirit.

IZAWA: I am not bragging or anything, but I think I was far

above in my tennis skills when I was young. But you are talking about the *spirit*, not the skill, right?

Izawa's experience was unique among nuclear power plant operators throughout the world. He told me that he hopes to be the last person ever to have to endure such an ordeal. He also hopes that with whatever time he has left, he can help people understand exactly what happened at Daiichi.

CHAPTER 7

HOW THE OTHER FUKUSHIMA SURVIVED

"Immediately after the quake, they subsisted on crackers;
after they got a water supply,
it was mainly rice."

—DAINI PLANT SUPERINTENDENT NAOHIRO MASUDA

At Daini, leadership would be the linchpin to the successes achieved. Although the circumstances were somewhat less dire than at Daiichi (for one thing, two electrical systems remained operable), there were numerous technical and leadership decisions to make. Neither of those remaining electrical systems was connected in a way that would safeguard all the reactors. Masuda understood that he'd have to organize his team of 200 to lay—by hand—five-and-a-half miles of high-voltage cable, in 220-yard sections (about the length of two football fields) weighing a ton each. And that was just the first step in navigating the disaster.

The decisions and actions of Naohiro Masuda in the hours and days following the earthquake and tsunami provide many lessons in extreme-crisis leadership. The first touches upon an answer to the question, *How does a leader deal with the fear and reluctance of unwilling followers?* Masuda understood how to enlist people to his cause and inspire them to keep moving forward despite their doubts and fears. The second involves what I call *using the brakes and the gas of leadership.* By knowing when to slow the recovery effort or speed it up, Masuda could keep his team working effectively and unwaveringly on this colossal task. In my opinion, it would not be an exaggeration to say that Masuda's leadership skills are what saved Daini from suffering the same fate as Daiichi.

DECISIVE MOMENT

We've been through the events of March 11 from several perspectives, but I will present the events of that day from yet one more because it powerfully conveys the point I am trying to make with this book. It's what I call "Team Masuda Leadership" at Daini. (My original inclination was to refer to this as "Masuda's Leadership," but that would run counter to Masuda's own humility and sense of fairness. This excellent leader has always shared the accomplishments of those dark days with his whole team, and I respect that quality tremendously.)

The earthquake that rattled Daini at 2:46 p.m. on March 11 was the biggest Masuda had ever experienced personally. From childhood, all Japanese are taught to duck under a table when they feel the earth move, but he'd never actually done it until that moment. The shaking was so severe in the administration building that he grabbed his hard hat and put it on.

Three to four minutes is an eternity when the ground is shuddering beneath you. Masuda spent this time working out the possible consequences in his mind. He figured that the reactors would shut down (scram). He shouted to the people around him to stay away from any hanging electrical cables. Once the initial shock abated, he announced that they would evacuate the administration building and await further instructions.

Normally, it was the job of the operations team in the control room to provide Masuda with ongoing information about the condition of the plant. Masuda assumed that he would be getting this information, and that his top priority should be to oversee the safe evacuation of everyone on his team.

Once he determined that everyone was uninjured and relatively calm, he walked over to the ERC. It was around 3:00 p.m. when he arrived. The response team leader reported that all four of the Daini reactors that had been in operation had shut down when the earthquake hit. They were all cooling down and working fine. It crossed Masuda's mind that perhaps they'd set a Guinness record when they stopped 4.4 million kW of energy at once; he was certain such a thing had never been done.

At about 3:10 p.m., an alert came up on the ERC television screen: A tsunami was coming but would probably be only about ten feet high. Masuda instructed his team to monitor it but wasn't that concerned about its height, even after a revised prediction calculated it at between twenty and thirty feet. He was preoccupied with other things that needed doing.

In Masuda's mind, a tsunami wasn't a violent, destructive force but a kind of slowly rising tide. In our 2013 interview, he told me that even if the report had predicted a wave height of sixty-five feet, he wouldn't have pictured it properly or been suitably alarmed. It was only after the power went out in the ERC that he understood the full potential of such an occurrence.

When the lights and instruments suddenly failed, Masuda realized that something had gone wrong. Reactor Units 1–4 and the ERC were forty feet above sea level, and the tsunami had reached all of them. The first floor of the ERC was flooded, which meant that the wave had crested as high as fifty-six feet—unimaginable.

Although the four reactors had been successfully shut down, the ERC started getting reports from the central control room that all the electrically powered reactor cooling systems were down. The emergency diesel generators had all failed as well, except one for Unit 3. Perhaps because it received its power from off-site, there was one electrical power supply still functioning at Daini, in a radioactive-waste building. Although far away from the reactors, it was a potential source of power.

For the moment, at least, the steam-driven turbine RCIC system for each reactor was running. Masuda knew that the reactors had to be cooled down, and that process required pumping systems powered by electricity. He needed to determine the extent of the damage and figure out which equipment they could fix. Despite the dangers of aftershocks and additional tsunamis, he had to send people out to make an assessment.

At the time of the earthquake, there were around 2,000 workers at the plant, including about 400 technical workers. Some 250 of these were official members of the emergency response team, and they gathered at the ERC. The remaining 150 evacuated to a baseball field nearby. Virtually no one left the site within the first hundred hours—and after the tsunami hit, those at the baseball field ran back to the ERC, making it crowded beyond capacity.

Masuda was not even sure if his people would be willing to go to the site if he asked them to, or whether they'd be safe if they went. On the whiteboard, he started recording the strength of each aftershock and the size and reach of each tsunami, to see if any patterns would emerge. The results did not seem encouraging, to him or the workers. Years later, he told me that, while the actual data may not have eased anyone's mind, it was his job to make sure they kept their hopes up and stayed to do their jobs.

ASSESSING THE DAMAGE

Shortly after 10:00 p.m.—more than six hours after the earthquake—Masuda gave the order to go to the reactor and turbine buildings and check on the cooling systems of all four reactors. Rather than naming specific workers to take on the responsibility, he asked his team leader to make the assignments—forty in all, divided into four teams of ten. Masuda had convinced himself that these people would be safe and, in turn, he convinced them. It was a vital mission if they were to restore the cooling systems.

The reactors had not yet sustained any core damage, so the workers would be safe from radiation, but there was a lot of debris to navigate in the dark. In retrospect, Masuda admitted that it wasn't the best time to send the men out, but he felt he had no choice.

Not one person refused to go, and he was grateful for their courage. He could see that they were afraid, but that they understood the importance of their mission. Before they left, he made it clear that they should come back immediately if he sent word. The tsunami alerts, which blared with every aftershock, added more anxiety to the situation. In all, there were more than 300 aftershocks that night. It was initially difficult for the various restoration teams to communicate—with the ERC, with one another, and with the outside world—as all mobile channels remained inoperable. One female engineer, Yukiko Ogawa, who was a team leader in the maintenance department and the most senior woman at Daiichi or Daini, was working to coordinate the flow of more than thirty restoration workers to and from the field safely.

Masuda knew he'd have some time before the workers reported in, so he focused on making sure the nonessential workers, and any pregnant women, evacuated to a safe location on the site. Then, he went to the ERC—which was in a seismic-isolated building—to see what was going on there. In a corner, he found several groups of women huddling together under blankets. Someone told Masuda that they weren't even TEPCO workers but subcontracted librarians. The tsunami had carried off their cars and, in any case, everyone had been told to stay where they were. Someone else might have simply moved on to the matters at hand, but Masuda couldn't do that. These women weren't meant to be at Daini, and he felt he had to get them home. Somehow, he found a bus and driver and asked him to drive the women as close to their homes as he could get, while exhibiting extreme caution in case of another tsunami. Some of the women made it home while others determined they'd be safer at the plant. In any case, Masuda had felt it important to take personal responsibility for them.

In addition to figuring out how to cool the reactors, Masuda's team had to ensure the survival of everyone remaining at the plant by finding and laying in supplies of food, water, and gas. None of them had been in the military, so they didn't have that kind of survival training. But he knew his people to be industrious and resourceful, and he was counting on them to rise to the challenges ahead. "One should never depend on subcontractors in that kind of emergency," he told me. "You need people who can do everything. That is the biggest lesson I learned from the accident."

It was after 1:00 a.m. when the inspection team Masuda had dispatched reported in with their assessment. Considering what they told him, he decided they should attempt to get at least one low-pressure core residual heat cooling system back in operation for each reactor. He asked an engineering team for input on where they should start, based on the pressure in each containment vessel.

From the outset, he shared all the information he was getting and all his plans with everyone involved, making sure they each had specific tasks. "If you let workers do whatever they think is best, there will always be redundancy," he told me. To provide the necessary guidelines for the work, Masuda wrote down his priorities on the whiteboard.

By dawn, the on-site restoration priorities had been fully mapped out. The team had figured out what kind of cable, motors, and other machines they might need and a procurement team was dispatched to obtain them. Fortunately, Masuda had a long history at the plant as well as an agile mind, and he could picture what was going on at the various sites from which his teams reported back. For that reason, he could provide concrete instructions. He never took notes, preferring to keep everything in his head.

NEW FEARS AND CONCERNS

Twenty-four hours after the quake, everyone at Daini was still working without knowing what had happened to their families and homes. Masuda knew that this was distracting them—perhaps even keeping them from listening fully to what he told them—so he tried hard to offer detailed instructions and often asked them to repeat what he had said. Sometimes, he had to shout at them, push those who weren't following his orders, but he did his utmost to remain calm and supportive of all who were fighting hard to get the job done.

During the first couple days, until the explosions at Fukushima Daiichi, Masuda and his team had little concern about radiation protection, but after that, it was uppermost in their thoughts. No one at Daini, including Masuda, had any idea how big the explosions had been, how far the plume might spread, or what was going to happen next.

When they saw the images on television and received reports of the destruction at their sister plant, Masuda ordered every site worker to evacuate to the ERC. When they'd assembled there, he could tell how nervous they were and struggled to figure out a way to lessen their anxiety. *What can I say?* he thought. *I know as little about the situation as they do.* Nevertheless, he reassured them, "I'll make sure no one is put in danger."

Masuda devised a restoration sequence for the four reactors. His evaluation team had at first determined that Unit 2 would be the first to reach its pressure limit, so that is where he decided they would begin. But during the preparations, the evaluation team corrected their calculations: Unit 1 would be the first to reach its design limit. Accordingly, Masuda changed course and determined they'd begin with Unit 1. He told everyone to listen up so he could explain the reason for the change. He knew that if they understood not just *what* to do but *why* they were doing it, they would make better decisions in the field.

COOL OR BUST

Essential equipment—in addition to what they had managed to assemble from the site—started arriving on the morning of March 13. Teams immediately started connecting cables and changing motors. Masuda was deeply concerned that if they tried to connect cables from the emergency diesel generator on Unit 3 to all four reactors, they might overload that power supply, which had been designed to support just one reactor. Masuda decided to instruct his team to connect cables from the radioactive-waste building, even though it was considerably farther away. They'd have to run five-and-a-half miles of extremely heavy high-voltage electrical cable by hand, but he believed they could do it if they all worked together.

The workers spaced themselves every six feet and began laying and connecting the cables. After a few cables were attached, Masuda realized how arduous this process was and changed his mind; they'd use the emergency supply at Unit 3 after all. The shorter route proved easier, though there was a different set of challenges to overcome. When they finally received a motor by truck, they had to figure out how to lower it to the ground without a forklift. Once they did get it down, they had to devise a way to navigate the tsunami debris to get it inside the building, then install it, and so on. Each step of the process involved creative thinking and backbreaking work. Masuda was constantly called on to advise, direct, and encourage his workers, and he rose to this challenge without fail.

For their part, the workers managed to stay positive and engaged. After connecting the first cable, they celebrated by applauding. After installing a motor or completing a test run, they openly rejoiced. And when a helicopter landed with equipment, they openly expressed their relief. Masuda continually thanked his men on the completion of each task, even if he'd yelled at them during it. In our interview, he told me that he believed these "thank-yous" had gone a long way toward keeping the workers motivated over the first few days of the recovery effort.

THEY SOLDIERED ON

Even as they worked feverishly to restore function in the cooling systems, the pressure in the reactors continued to rise. Time was not on their side, and they knew it. They calculated that they had just two days before the point where they'd have to vent the containment vessels (the same drastic condition their counterparts at Daiichi were already facing).

The operations team attempted to stall the pressure buildup by spraying water inside the containment vessels and Suppression Pools. They also circulated the water already in the containments. In the end, on March 15, they restored the cooling functions just two hours before the pressure reached critical levels. All 2,000 workers at the site had worked together, each providing his expertise, to accomplish this Herculean task. And Masuda had presided masterfully over the operation, maintaining a cool head and firm hand.

He shared with me that he'd heard very few complaints. For the most part, the workers had seemed to trust their leaders implicitly.

A few of those working to connect the cables did approach him to say they wanted desperately to see their families. He wanted them to stick with it at least until the 23rd, allowing plenty of time to cool down the reactors. He asked them to stay until the reactors reached a certain point of safety. He was reluctant to provide them with a date, giving them a goal to work toward instead. He knew that if even a few people left the site, the morale and focus of the others would be affected.

THE HUMAN ELEMENT

After restoring the cooling function, Masuda decided to reorganize the shifts so that people could go home and stay for a few days at a time. Starting April 1, he told his workers he'd need only a few people on the site. Everyone else was free to take some much-needed time off.

"I wished I could have let them go earlier," he told me later. "It was the beginning of the school year, and they wanted to be with their kids. Under all but the most critical circumstances, I believe that family should be a priority."

By the third day, it was possible to get phone calls out, using the internal phone system via Tokyo. Masuda worked hard to arrange time and access for each of the hundreds of workers to make phone calls to their families.

Another issue was that the female workers, numbering twenty or more, had no access to showers for a week. Masuda changed his socks for the first time a week after the disaster and didn't remember washing his hands for ten days. Almost immediately, he recognized the necessity for a women's restroom as well as workable toilets, but they had no water supply. Ultimately, Masuda arranged for a truck to bring water for drinking and sewage purposes and to make showers available for all, but this took time—and when it did arrive, the water was ice cold.

Masuda allowed daily shower time for the women while restricting the men to once every four days. One female worker, an architect in her twenties, had concerns about the water being pumped through the pipes, claiming that it might be contaminated. Masuda understood her point and made a mental note about filtering the water, or at least testing it for radiation contamination. Here again are the makings of successful leadership in a crisis: valuing the advice of experts in their fields, even if impractical, and giving their suggestions consideration. If someone suggested a different way to perform a task, Masuda considered it carefully. He knew that others might have knowledge about the reactors that he didn't possess and was careful to honor their expertise. He held a meeting each morning and night to share information of all kinds, about concerns large and small, including issues of personal hygiene, safety, and forecasts.

In addition to improving their work hours, Masuda wanted to make sure his workers were eating properly. After the quake, they had subsisted on crackers; once they'd gotten a water supply, it was mainly rice. In May, the chamber of commerce of Iwaki City offered *bento*[1] delivery service to the weary employees, and Masuda was very grateful—but he worried that his people would become accustomed to the service. He decided it would be better to improve the situation little by little. For the first two days, there were dinner deliveries only; then both lunch and dinner were brought in, and then breakfast. In that way, the workers were made to feel that conditions were improving little by little, and they never took these small boons for granted. He insisted that each employee wash his own *bento* box—and of course, he did so as well.

Masuda believed that getting back to normal routines was essential if they were ever to feel that progress was being made through the crisis. Things such as morning exercise and keeping up daily routines were an essential part of this, he reasoned, so he turned his attention to these things, in addition to everything else. He requested that members of Daini's female soccer team lead a vigorous morning exercise routine (though he found himself quite physically challenged after just a few minutes on the first day of exercise).

Masuda continued to focus on the widest range of things in the days following the disaster. He decided that working, eating, and sleeping in the same space wasn't healthy, so he found an unused building near the ERC and set it up as a place to eat and rest. That way, his staff had to take a walk occasionally and spend some time in a different environment. On occasion, instead of returning to the ERC after dinner, people simply bedded down for the night.

More than half of Masuda's 400 workers lived in Fukushima Prefecture, in the nearby environs of the plant. About twenty-three of them lost their homes, including Yukiko Ogawa, the maintenance team leader who lived and worked at the site for two-and-a-half months. She could not return to her home in Tomioka Village for more than a year, and when she did, she found all of her possessions gone. Eight workers lost family members in the tsunami.

A psychiatrist from the National Defense Medical College, which is managed by the Ministry of Defense, had started showing up at the plant soon after the earthquake, observing mental health conditions there as well as at Daiichi. The psychiatrist confided to Masuda that his workers were suffering from severe post-traumatic stress disorder. In 2013, Masuda said that he'd found it very helpful to have the psychiatrist on site—as important as having medical doctors there to take care of the physical injuries.

OBSERVING MASUDA

During the year that I spent in Japan after the disaster, I gained tremendous insight into extreme-crisis leadership. Much of this insight came directly from my observations and conversations with Masuda and others about the events at Daiichi and Daini. I'd understood that leading under extreme conditions involves a unique set of skills—but there, I saw this manifested in action. Fear of imminent death can hamper decision-making ability on the part of the leader as well as a team's willingness to follow orders. But this fear can also bind a unit together, even to the point of defiance of outside authority. (This was exemplified at Daiichi, when Yoshida defied the order from headquarters, under the harshest duress, to stop the seawater injection.)

As I said in Chapter 2, the leadership abilities and techniques of those in charge are second only to the cause of the event itself in determining the outcome of a disaster. Masuda exemplifies this conclusion: His understanding that fear and uncertainty can be ameliorated by providing precise, digestible information; his innate ability to know when to apply the gas and when to apply the brake; his competence, which earned the respect of his workers; and his ability to make sense of the seemingly unfathomable—these things made him the consummate leader for the situation. Add to those qualities his foresight in preparing for a long and sustained effort under the most arduous conditions. Under Masuda's leadership, by mid-March—on the same day that the Unit 4 Reactor Building exploded at Daiichi—the reactors at Daini were safely being placed in cold shutdown.

Masuda stayed on at Daini for several years after the crisis had passed, until he was assigned to lead the decommission of Daiichi's reactors.

Figure 5. The author (left) at Daiichi (source: TEPCO)

CHAPTER 8

OUR ARRIVAL

We were running and thinking at the same time. As a crisis leader, I knew that I had to get this situation organized.

When the team from the United States arrived in Tokyo, we were immediately facing countless logistical and leadership challenges. We had no data. The conditions at the accident sites were deteriorating fast. The crisis was cascading faster than we could respond to it. We had to deal with life-and-death recommendations and get organized at the same time. It seemed that every minute of the experience was providing a new leadership lesson.

As the plane arrived at the terminal, I was called to the front of the airplane where two people—one security person and one U.S. Embassy staffer—met me and escorted me through customs and immigration. We didn't stop; we just sailed right through. Then they drove me into Tokyo and directly to the embassy.

This was my first time in Japan. From my window as we cruised along the "wrong" side of the expressway, I noticed right away that everything seemed clean and orderly. The expressway was landscaped perfectly, in contrast to the video I had just seen of the devastation in Sendai. I wondered how it was possible that so much damage from the earthquake and tsunami had occurred just miles from me, yet this area was untouched. I thought about the devastation, the thousands of people dead or missing. I prayed. After about an hour's drive, we arrived at the embassy downtown, and the first thing I wanted to do was go to the ambassador's office. Matt Fuller, the first of the ambassador's immediate staff I met, greeted me and handed me a copy of an email. "Have you seen this?" he asked.

I hadn't. I looked at the note as Matt informed me that we were about to go into a video conference with some admirals. About thirty seconds later, we entered the conference room, where I met Ambassador John Roos for the first time. Some officials, including some high-powered admirals, back in Washington were insisting that the Japanese should conduct a broad evacuation, that they should get water into the spent-fuel pools and reactors immediately and take heroic measures to "stop the event." Contrary to their well-intentioned input, I knew that these "heroic measures" would mean exposing innumerable people to lethal doses of radiation.

One of the NRC engineers was sitting with me and responded very negatively to those comments. He said, essentially, "Every idiot in the world knows that they need to get water on those reactors." Shocked at where the conversation was going, I tried to guide it or, better yet, shut it down by asking specific questions of the officials and suggesting that our staff work with their staff to iron out any misapprehensions and disagreements. What I meant was that we should work together on specifics to come up with some fact-based recommendations for the Japanese government. My goal was to prevent the engineer from digging us a bigger hole. The conversation ended. The engineer and I walked into the hallway. "I don't think that was an appropriate discussion with the brass who were on that call," I told him.

"Look, admirals are a dime a dozen," he replied.

Shocked all over again, I said, "One of those admirals runs the nuclear navy. He has many nuclear reactors, and if you keep it up, I know exactly where he will put one of them." I understood how important it was for us to show respect to the military in this context, in spite of the drastic and rather aggressive points they were making.

That meeting was my first interaction with Ambassador Roos and Matt Fuller. I knew instantly that they were both intelligent and passionately engaged in finding solutions, and that they were seeking a "nuclear person" who could help them understand the event. Ambassador Roos is a highly successful Silicon Valley attorney and his wife, Susie Roos, is equally as sharp and successful. She is an extremely successful lawyer in her own right, and is very inquisitive—she was particularly interested in the fate of Daini. She kept asking us, "What about Daini?" I will never forget the graciousness of Susie and Ambassador Roos. I came to consider them exceptionally great friends and still do.

Someone told me during those first few minutes after my arrival that the Unit 4 reactor building at Daiichi had just exploded. That was the third major explosion. Units 2, 5, and 6 remained intact, but we knew that Unit 2 had sustained significant fuel damage. Because of the explosion at Unit 3, a piece of debris had punctured a hole in the side of Unit 2 and, as luck would have it, that hole was allowing the hydrogen from the reaction of the melting fuel to escape. Otherwise, that building would surely have exploded as well. Much of the debris that resulted from the explosions of each reactor building fell into the spent-fuel pool, rendering the condition of the spent fuel unknowable. And if that was not enough, there was profound concern about potential spent-fuel fires that would release massive amounts of radioactivity into the atmosphere.

Just before I joined our team in the conference room, I stood in the hallway thinking about the situation. I took a second to add the Unit 4 explosion into my assessment of a situation that was genuinely beyond my imagination. The earthquake and tsunami and thousands of continuing aftershocks had caused severe damage to all the reactors, and there had been three major explosions already. Severe reactor core damage was inescapable.

The operators at Fukushima Daiichi were facing a daunting array of technical issues. As advisors, we struggled to help the Japanese address them. There was no electrical power and, consequently, insufficient injection of water flowing into the reactors. This was a surefire recipe for reactor core damage. Radioactive plumes necessitated the evacuation of more than 163,000 residents. The reactors were essentially becoming evaporators, steaming off the seawater that was being injected from firetrucks and leaving ocean salt behind. That put us in unknown territory in terms of dealing with reactor operation: We didn't know what that salt would do to water flow in the reactor cores, nor its effect on the metal of the reactor vessels.

Separately, there was worry that the reactors might go re-critical (begin the nuclear reaction again), releasing untold radioactivity and—even more dangerous—neutron radiation. The list of serious concerns went on: the molten reactor fuel could penetrate the bottom reactor vessels (think of the blood from the creatures of the *Alien* film series searing right through metal), resulting in a steam explosion that would release massive amounts of radiation. Thermal images seemed to indicate immense problems in the cores, containments, and spent-fuel pools—though we would soon learn that we didn't really understand what those thermal images were telling us.

As I mentioned, some American officials were pushing for the Japanese to take "heroic measures" to stop the accident. There was a rumor circulating that the TEPCO operators would evacuate the site, leaving it in the hands of the Japanese Self-Defense Forces (SDF). Already, the American anti-nuclear crowd was posting fraudulent YouTube videos that were unnecessarily scaring the Japanese as well as embassy dependents. Countless foreigners were trying to get out of Japan, while the Japanese attempted to cope with the earthquake and tsunami.

Some embassies in Tokyo were evacuating. An American aircraft carrier was sailing around in the radioactive plume. We had 153,000 resident Americans plus tourists in Japan to protect, including all the frightened dependents of embassy workers. We were not sure that there were enough potassium iodine (KI) pills available to protect people from radiation. "Radiophobia" spread across the world, propelled by bogus reports of elevated radiation levels around Japan. Everyone was concerned with the air and water supply in Tokyo. There was a discussion about the possibility of having to evacuate Tokyo completely. The press was everywhere, including in our hotel. The relentless need to respond to the press was getting in the way of our response to the disaster itself. At one point, a Japanese politician said to me, "We are managing for worldwide protection

from this accident, and we need to move from peace to a war foot-
ing." An NRC commissioner called the accident "biblical." Although
overstated, those comments felt accurate at the time.

On top of all those issues and more, communications and data
flow were in utter disarray. Either there were no facts, or there were
wrong "facts," and misinformation was flowing much faster than
accurate information. Understandably, the Japanese were reluctant
to share what they knew, partly because they didn't always understand the implications of what they were hearing. Neither did we.
We were running and thinking at the same time. There were more
unknowns than knowns regarding the Japanese response, and the
American response was just starting to get underway.

Figure 6. Team NRC first weeks

CHAPTER 9

HEROIC ACTS

*They suggested that they would not let the workers leave
the site. I understood what that meant.*

As I mentioned in Chapter 8, there were some American leaders who insisted that the Japanese take heroic action in response to the accident—meaning that operators would knowingly be subjected to lethal doses of radiation. It would be difficult (to say the least) for me to forward this recommendation. It was a leadership dilemma unlike any I had ever faced.

As a leader on the ground, I understood that, while the situation was grave, no one was dying of radiation exposure—at least not yet. I think some of our leaders back in the States were remembering the heroic firefighters and helicopter pilots who sacrificed themselves at Chernobyl. They believed this course of action to be necessary at Fukushima Daiichi. I firmly believed otherwise; thus, my predicament.

I had several choices: Pass my superiors' recommendation on to the Japanese immediately; slow-walk it; or just not deliver that message at all. I chose to raise the issue subtly with the Japanese leaders, while simultaneously providing my own recommendation not to order such heroic acts. Why? In addition to my theory that the worst had already happened, I understood that the Japanese were struggling simply to keep people on-site and working on a remedy. It seemed to me inappropriate for a U.S. government official to step in and say, "You must do more, including asking plant workers to sacrifice their lives." There were enough people suffering from the earthquake and tsunami already; I saw no need to demand more lives.

The officials in Washington, some 7,000 miles away, had a distorted view of how to run the event. Clearly, we had not learned as much as I would have hoped from the experience of Three Mile Island—namely, that you shouldn't and really can't run a crisis from a distant headquarters. These situations must be directed by experienced leaders operating as close to ground zero as possible.

While I was still trying to get my bearings at the embassy, Goshi Hosono, the special assistant to the Japanese prime minister, called for a meeting. Hosono, Ambassador Roos, and I, along with other members of the Japanese government, met in a small garden conference room in the Hotel Okura. Due to the power shortages, the conference room was poorly lit and extremely hot for March. We were there to discuss the rumor that the Fukushima Daiichi workers might evacuate the plant. I knew that the Japanese were getting suggestions from Washington about the "heroic act" issue and I very much wanted to discuss it.

At that point, TEPCO was allegedly contemplating leaving the site. The Japanese Self-Defense Forces (SDF) had stopped working because of injuries sustained when the Unit 3 reactor building exploded; they were reluctant to do more work around the reactors without further guidance. Prime Minister Kan had visited TEPCO Headquarters[1] earlier that day and directly ordered those in charge there to keep the workers at Daiichi. I understood that to mean that the workers would be staying in place.

During that meeting, I subtly made it clear that I was not in favor of "heroic measures" on the part of Japan's SDF or the Daiichi operators. The plumes of radioactivity that were escaping the Daiichi reactors represented an increased risk of cancer in the future, but I doubted that either plant would release radiation outside the boundaries of Daiichi at such levels as to cause immediate death from overexposure. This had happened at Chernobyl only because the fuel was physically ejected from the core and exposed to the atmosphere.

I also believed that the effort the operators were making to inject seawater was working, and that conditions were getting less dangerous, not more so. My position was that only if conditions degraded further should we recommend more aggressive action. As we spoke, the reactors had already gone days without human intervention, and the radioactive plume had not gotten bigger. It seemed to me that the accident had progressed as far as it was going to, in terms of off-site radiation exposure. I believed that after the Unit 4 reactor building exploded, the worst was over, though much effort would be needed to stabilize the reactors. It was appropriate for us to monitor what was going on and challenge plans and ideas as we saw fit—and to continue to encourage the operators to keep working—but not to request or demand any "heroic acts."

As an extreme-crisis leader, I had to trust my own sensibilities and judgment beyond those of my distant superiors, just as Yoshida had done regarding the seawater injection. Later, when I was asked about Yoshida's action by CBS News' Scott Pelley[2]—who termed Yoshida's act of defiance "not very Japanese-like"—I responded that in those moments, Yoshida wasn't Japanese; he was a nuclear professional, and his act was "operator-like."

This situation reinforced for me the value of the U.S. protocol for licensing reactor operators. Our process licenses and empowers operators to make appropriate technical decisions without direction from the White House or NRC leadership. In short, *we allow our operators to do their jobs.* I've heard from some Americans that the situation in Japan would never exist in U.S. nuclear plants, because we understand the importance of the operator's prerogative and would never overrule an on-site operator's decision. I hope they're correct about that, but the truth is, I'm not so sure our leaders would act much differently from the Japanese.

During our meeting in the Hotel Okura, when the Japanese raised the need for helicopter drops of water on the spent-fuel pools, I told them that I thought this would be ineffective and could result in unnecessary radiation exposure to helicopter crews. Soon, this issue would dominate all others.

CHAPTER 10

PUMP UP THE VOLUME

---

*"A cup of water applied at the right time
can put out a forest fire."*
—John Abraham

O ur team believed that the use of firetrucks to feed water to the reactors was an appropriate but short-term solution. As crisis leaders, we knew that if another earthquake and tsunami struck, conditions could change dramatically. To advance the response ahead of the events on the ground, we felt the pressure to start working on a longer-term solution. In the absence of any long-term solution proposals by the Japanese, our NRC team set about working with them to improve the robustness of the water systems that would supply cooling water to the reactors and spent-fuel pools.

It did appear to us at times early on that the Japanese were reluctant to accept help from us. After the explosions of Units 3 and 4 and the resulting first-responder injuries, the operators on the ground and the Self-Defense Forces seemed to be frozen in inaction. We needed to encourage the Japanese to act, despite the fear that was causing a sort of inertia among many of them. We had to propose actions they could take without risking further injury or causing a catastrophic event. Our development of a pumping system unintentionally achieved this end by demonstrating that we might act on our own if they did nothing. This seemed to stir the Japanese, who were somewhat threatened by our interventions.

Although their water drops would not prove particularly effective, they were a way the Japanese could demonstrate that they were, in fact, taking decisive and even heroic action, much as the Chernobyl pilots had. By showing the helicopters on television, they hoped to reassure their people that their government was working hard to improve the situation.

There were many reasons that the Japanese were reluctant to accept American help, including political/diplomatic considerations, technical/operational issues, and competitive forces. From a political/diplomatic perspective, the U.S.-Japan alliance is a very sensitive issue in Japan. While the Japanese support the alliance, there is some concern by politicians that America will always push its will on the Japanese military. Some in the Japanese government who did not understand nuclear technology feared that the Americans might impose themselves technically, especially because the reactors were American General Electric facilities. TEPCO itself was not interested in being taken over by American nuclear experts or the American government. For these reasons and more, there was significant reluctance to allow an American-designed-and-built pumping system to be implemented at Fukushima Daiichi.

At the same time, our NRC team continued to worry about the efficacy of using firetrucks to supply water to damaged reactors and pump the water into the reactors' spent-fuel pools. One of our senior reactor analysts told me that to decrease the risk of damage significantly, the Japanese needed to install additional water injection points into the reactors wherever possible, to increase the diversity of their sources. I factored in his advice as we formulated plans, coming up with a strategy that was infinitely more productive than just adding more pumps. The Japanese thought that merely adding pumps to inject through one point would improve the reliability of the connection, but we understood that it would not. Our experts were encouraging the establishment of more connections, or injection points. These would improve the redundancy and reliability of the pumping system and help guard against disaster in the face of another seismic event.

The Daiichi operators had connected firetrucks to introduce

seawater to the reactors, but fire pumps are not designed for this kind of long-term pumping—nor are they designed to withstand earthquakes and tsunamis. Certainly, we thought, a hard pipe injection system would be more reliable. To improve the probability of successfully cooling the reactors long-term, they would have to switch away from the firetrucks altogether. We began to push the Japanese government and TEPCO to enhance the reliability of the pumping systems, envisioning a system where water was conveyed to the reactors and pools through large, hard piping. Our intention was to make the system so rugged that there would be no need for workers to expose themselves to lethal doses of radiation while arresting deterioration of the reactors and spent-fuel pools.

Our team and TEPCO engineers jointly designed the pumping system. I am not sure whether they were coordinating with us as a courtesy, or if their leaders were even aware that their engineers were cooperating. In fact, one night the joint team worked on the design until 4:00 a.m., in a conference room at Hotel Okura. We provided the resulting plans to a contractor in Europe, who told us they could indeed provide such a system and offered to do it pro bono. (Ultimately, however, that offer went from pro bono to a cost of tens of thousands, and eventually millions of dollars.) The contractors figured they could assemble the pumping system in Australia and fly it via Air Force Boeing C-17s to Japan, but this became very difficult logistically and very thorny politically.

We found out the real cost of constructing the pumping system after the first "train" (of four) was built. The original estimate of $750,000 ultimately ballooned into $9.6 million, for all trains. As a result, we had only one train constructed, made up of three pumps and associated piping and valves. Our NRC engineers determined that this one system could supply all the reactors and spent-fuel pools, so we canceled the order for the other trains, saving millions of dollars and shortening the time from concept to completion. Theoretically, this also lessened the potential for diplomatic and political pushback.

Several times as we worked on coordinating the arrival of the pumping system, the State Department (DoS) issued "stand-down" orders. This was extremely frustrating because we knew how much the system was needed, but it seemed that the Japanese government (and, to some degree, our own DoS), was overly cautious about the operation. In the end, though, the pumping system was flown to a U.S. Air Force base in Japan, and we sent one of our engineers to receive and sign for it. This was humorous, because he was signing for a $750,000 project that had never been officially approved by anyone. I told him not to fret about it, and that if anybody went to jail over it, it would be me.

In a call to Prime Minister Kan in the immediate aftermath of the event, President Obama had said that the United States would do anything necessary to support the efforts and had offered a blank check along with our experts (including me) and U.S. AID (Agency for International Development) for the first thirty days.

U.S. AID was traditionally mandated to help Second- and Third-World countries, not a First-World country like Japan—but there is a first time for everything. In that first thirty days, a bilateral arrangement was put in place, that I'll discuss later, to continue our work after the termination of U.S. AID support.

During an unfolding crisis, you can eliminate bureaucracy, and early on, there was little of it in our government's response. U.S. AID did a fantastic job of getting us to Japan, where we set up an independent Internet LAN in the U.S. Embassy. The government suspended many rules, regulations, and policies. People understood that we might need to break a few rules and take swift action in the early stages, but, as often happens in an ongoing crisis, the bureaucratic burdens tend to reassert themselves as time goes on. On top of that, special interest groups, albeit for good purpose, began to try to impose restrictions on activities.

Once the pumping equipment was on Japanese soil, we had to get it to Daiichi. We sought permission to transport the pumping system from southern Japan up to Fukushima Daiichi, about 185 miles, which is at least a four-hour drive. We knew the clock was ticking on saving the reactor cores and spent-fuel pools, but the negotiation ate up several precious days, as we ran into the sort of bureaucratic problems that had hindered us every step of the way. Now, local government officials were hesitant to approve the transport along Japanese roads. There were discussions about the required permits and whose vehicles would tow the equipment, American or Japanese. It quickly became clear that they would have to be Japanese. The Navy supplied fresh water barges, but the Japanese insisted that their tugs push them northward to Daiichi. Meanwhile, thousands of aftershocks threatened the firetrucks injecting water into the reactors.

We realized that we also needed some specialized equipment to go along with the pumping system—spray nozzles to direct the water, among other things—so we made a list and instructed our procurement offices to find the equipment in Japan so as not to unnecessarily fray relations. As they searched, they learned that the Tokyo Metropolitan Fire Department had purchased the same items. It became clear that the Tokyo Metropolitan Fire Department was building a duplicate system, a fact that came as quite a surprise to us.

Ultimately it could be said that by being proactive—and maybe even provocative—we spurred the Japanese government to act on its own. As I understand it, in the end, our system was transported to a port near Daiichi but never did make it to the site. Someone told me that just one of our pumps was ever used.

The State Department in Washington was a hindrance to building and transporting the pumping system, for reasons that remain unclear to me. I suspect that the Japanese were putting pressure on them to kill the initiative—either that, or Washington was afraid that our little band of warriors was trying to take over TEPCO. Or maybe they just didn't understand the need for the injection system. In the end, Washington criticized us for building something that ended up not being "necessary." The best we could offer in our defense was the fact that building it had catalyzed the Japanese government to build its own system.

BABY STEPS ARE NOT ENOUGH

That first week, a Japanese leader told me that there was much debate between the Ministry of Defense (MOD) and the Ministry of Economy, Trade, and Industry (METI) over the configuration of the nozzle that was to be attached to the pumping system. The question was whether it should be a spray nozzle that would spray water completely over the spent-fuel pool, or a hard pipe that would fill the pool from the bottom. I don't think it mattered which one it was; the controversy was actually over which of these entities knew the most about the reactor and spent-fuel pool designs. It was, quite simply, a proxy power struggle over control of the accident between civilians and the military.

NISA and TEPCO reported to the government that the MOD did not understand the spent-fuel pools and that a spray nozzle would be most effective in keeping down the radiation plume. The spray nozzle was the correct choice. After that debate, we no longer went to MOD for our meetings, but conducted them at the offices of METI/NISA and TEPCO. MOD had lost the battle for control in an internal struggle that I believe was technically baseless.

As a crisis leader, I was most emphatically not concerned with the internal politics related to the pumping system. Nothing is more crucial for an extreme-crisis leader than maintaining focus on getting and keeping ahead of the accident. Allowing a crisis to cascade can quickly overwhelm any planned response. Once we are outside of a planned response, the probability of a successful outcome becomes exponentially more remote. Political maneuvering often arises during an extreme crisis, and political considerations can overrule technical ones. My career leadership roles, my time on Capitol Hill and in the NRC chairman's office, and my many years in the field have helped me understand that politics must be managed just as the technical issues are. A crisis leader must stay focused on the mission, but politics cannot be disregarded. In Japan, I served as both a political influencer and a situational/technical interpreter.

With General Douglas MacArthur's quote[1] about "too late" echoing in my mind, I realized that baby steps would not be sufficient in this situation. We made that mistake in our response to Hurricane Katrina and as far back as the Great 1906 San Francisco Earthquake and fire. MacArthur was very much a political operative, not only masterminding the war but overseeing the rebuilding of Japan in peacetime. Most important is the leader's continued effort to keep the speed of the response ahead of the events. John Abraham, a friend of mine, once told me that "a cup of water applied at the right time can put out a forest fire." During the Fukushima Daiichi accident response, we needed much more than a cup, and we needed it *immediately*.

CHAPTER 11

STARRY, STARRY DAY

"Hey, Chuck's a great American, but an operational
constraint limits the amount of information he can get."
—TROY MUELLER

The day after the disastrous video conference with Washington officials, I was summoned to the ambassador's conference room with three flag officers, who were wearing a total of eleven stars: Admiral Robert Willard, Commander U.S. Pacific Command, Admiral Patrick Walsh, Commander of the Pacific Fleet, and Burton Field, Commander, U.S. Forces Japan, and Commander, 5th Air Force, Pacific Air Forces, Yokota Air Base. These fine gentlemen were clearly anxious to protect our military personnel. We explored the issues for two-and-a-half hours of intense discussion. Apparently, they had been under-impressed by the video conference the day before and had come in person to get answers. (Do I need to tell you that the first two days were extremely intense and a little rocky for me?) These men were deeply interested in learning about the degree of penetration we had into the Japanese government, as well as the operations at Daiichi and Daini.

I didn't grasp then that the driving force of their concern was that our aircraft carrier, the USS *Ronald Reagan*—along with 16,000 crew on twenty support ships—was sailing through plumes of radiation in the Pacific. Further, by this time, the USS *George Washington*, ported in Yokosuka, had monitored relatively high readings and was considering its options.

The flag officers asked me for predictions about future radioactive plumes, what they might look like, and where they might be. They made it clear that the *Reagan* had an obligation, as part of the U.S.-Japan alliance, to protect Japan from threats, and that they were also mandated to help with the relief effort. My first (silly) reaction was, *Surely you guys can find a place to park that thing somewhere, or could create one with the armament on that fleet!* But that sentiment would have added nothing to this important discussion. As a leader, I had to expand my view to consider the other stakeholders, whose interests far outweighed my own.

Besides the issue of penetration, the military guys asked some very tough questions that I'll never forget, including grilling me about the integrity of our information. I was sweating throughout the meeting, because I had few answers to offer and didn't want to come right out and say, *The NRC gave me orders not to penetrate TEPCO, but to stick exclusively with NISA and the government of Japan.* Of course, this order was not much help in gaining clarity on the accident. It handcuffed our team. It made working with the U.S. military delicate, to say the least, because they have deep penetration into Japan's Self-Defense Forces. We were in a "bridging situation," attempting to appease military requests while following NRC protocols.

Because the U.S. Nuclear Regulatory Commission is a government agency, up to the time of the accident, it worked primarily with NISA and had little working relationship with TEPCO. But the American nuclear industry had some links to TEPCO. We immediately recognized that NISA was a dysfunctional organization, but the NRC was committed to working with them. Ultimately, we and the private nuclear sector became the de facto liaison with TEPCO.

Through those discussions with the flag officers, I learned that among the vulnerable groups on the aircraft carrier and its escorts were women of childbearing age, pregnant women, and lactating women. Imagine my surprise at learning that some women in military service pump their breast milk and fly it back to their babies! As I learned more, I grew to appreciate the level of concern of the men I was dealing with, but I honestly had little information to impart.

We talked about the design of the plant and the generic reactor accident scenario, but regarding specific conditions, I didn't have the goods. And, since the accident had blown the instrumentation at Daiichi across the landscape, I wasn't alone. After two-and-a-half hours, I still hadn't managed to say, "Look, this really isn't my job." Looking back, I think maybe they would have appreciated hearing that, but at the time, I just didn't feel it would help matters.

Fortunately for me, Troy Mueller, the Naval Reactors[1] Program director, spoke up and said, "Hey, Chuck's a great American, but an operational constraint limits the amount of information he can get." Not that this improved my opinion of myself at that moment.

As an extreme-crisis leader, this experience was instructive to me in two ways. First, it opened my eyes to consequences of the accident that had been beyond my vision. In cases of extreme events, the scope of one's vision is necessarily limited. Hearing from the military reinforced for me that even someone with experience, expertise, and familiarity can overlook important issues without input from others. My philosophy is to seek out a broader vision whenever possible. Second, I understood the consequences of the limits in the rules of engagement that the NRC had placed on us.

I was challenged in this meeting to prove my credibility, and my limited scope made me feel incompetent for a while. I didn't understand at that point why the NRC had put those constraints on us. Of course, there's a big difference between the NRC penetrating a U.S. utility and the American government penetrating a private Japanese company. Today, TEPCO maintains good relationships with the American nuclear industry through the Institute of Nuclear Power Operations (INPO)[2] and its international counterpart, the World Association of Nuclear Operators (WANO).[3]

EVACUATION DILEMMA

Immediately after the accident, we held discussions on how best to protect American citizens in Japan, including military personnel and tourists. In debates about an evacuation, we grappled with the uncertainty of the data surrounding reactor conditions and radioactive releases. In the years since then, there's been endless discussion about the decisions made on evacuation—and everything else, for that matter.

From the military perspective, a broad evacuation was necessary.[4] I think this perspective was born from the naval reactors culture, where any radioactive release is bad. According to NRC guidelines, our normal recommended evacuation radius in the United States would be ten miles. In this case, because we didn't own the reactors and didn't know the extent of damage, the NRC was willing to go somewhat further. Complicating matters was the fact that there were several competing radiation plume computer models. Japan had its "System for Prediction of Environmental Emergency Dose Information" (SPEEDI),[5] and the United States had its own models. There were a lot of problems with SPEEDI. Many Japanese officials didn't even know it existed, and most local officials couldn't understand it.

In any case, the loss of electrical power had caused failures of the system with wide-ranging and unpredictable consequences. Because of these issues and others, for example, poor evacuation plans, poor public communications, and confusion in the crisis response, the Japanese government and its systems were not considered highly credible.

One U.S. computer model was designed specifically to measure the radioactive releases in the immediate aftermath of an accident, while the other maps out longer-term effects; but neither of them could reliably model the complexity of the Fukushima Daiichi accident. We just didn't have enough good data to get reliable numbers. As a result, the United States made some big assumptions that provided some questionable outputs.

Because of its experiences with Three Mile Island and Chernobyl, the government developed a "radioactive plume" software model to predict the damage that would come from *one* compromised reactor core. The accident we were dealing with might end up involving multiple reactor cores and up to seven spent-fuel pools. We just didn't know, so we couldn't know how the model would relate to the situation. Ultimately, a decision was made to model one-and-a-half damaged cores as a "super-core," even though multiple cores had been damaged. Modeling more than this might have resulted in misleading results. It was still all guesswork.

The U.S. government has no authority to order an evacuation in Japan or any other nation. They can only advise. What followed was not an evacuation order per se (that was something imposed by local governments within their zones), but rather a strongly worded advisory or "warden's warning" issued by the U.S. State Department on March 17. It was meant to encourage U.S. citizens to leave Japan or, at the very least, stay out of the danger zone in Fukushima Prefecture (see Figure 7). The embassy staff told us that there were very few Americans in that area, perhaps fewer than 300. Most of them were married to Japanese nationals and would follow Japanese guidance. The U.S. decision to advise an evacuation had two purposes: to provide the U.S. military with guidance on where military personnel could safely travel, and to send a message to the effect of *Don't become part of the problem by going into these areas while the Japanese attempt to evacuate.* The more official basis of the advisory came from data gathered by the Department of Energy, using its Atmospheric Monitoring Center radiation surveys. Mixed into the decision-making were concerns for above-background levels of radiation as far away as the Yokosuka military base, future spent-fuel fires, and the deteriorating condition of the plant.[6]

The Japanese started with a small evacuation zone, but kept expanding and altering it based on radiation readings. The zone was poorly delineated and the policy somewhat incoherent. Each time the zone changed, public confidence in the government declined. Although the SPEEDI system modeled a plume being blown toward the northwest, the small villages of Namie and Tomioka, near the nuclear plant, were evacuated right into its path. The 8,000 residents of Namie were first evacuated to Tsushima, a city of 1,400 residents 17 miles inland. Firefighters would have to dig holes in the ground for the people to use as toilets. Local streams were used as a water source to make rice balls.

After the Unit 1 reactor building explosion, these people would be evacuated again—to the great frustration of Namie Mayor Tamotsu Baba—and the area would become one of the most radioactive in the disaster. Some in the Japanese government withheld SPEEDI information from local mayors—including predications that could have prevented mass contamination of citizens. This caused considerable animosity among citizens in the region and has subsequently led to lawsuits over radiation exposure to these communities.

The Americans were much more conservative. In part because of the paucity of data, we expanded our evacuation zone recommendation beyond the 30 kilometers (18 miles) that the Japanese had established. In the end, the NRC recommended a 50-mile (80-kilometer) evacuation zone that included three large cities. The Japanese settled on a modified 18-mile (30-kilometer) evacuation zone. The differing advice generated significant media and political attention. The Japanese press concluded that the Americans were more focused on protecting people than the Japanese government, sowing seeds for a loss of trust in the Japanese government by its people.

In some ways, this reminds me of a lesson I learned from Governor Richard Thornburgh of Pennsylvania in his response to the accident at Three Mile Island. The NRC had told him that with a reactor core meltdown, there was a potential for hydrogen buildup in the reactors, which might lead to an explosion (exactly what would later occur at Daiichi). They discussed the potential for evacuation. Thornburgh consulted with his local emergency managers and learned that the county north of the plant would evacuate across the bridge to the county south of the plant, while the county south of the plant would evacuate to the north—across the same bridge! This created a rather absurd conundrum.

When faced with this uncertainty—a hydrogen explosion versus contradictory evacuation routes—Governor Thornburgh went with the facts he had and did not order a full-scale evacuation. As it turned out, he made the right decision. Governor Thornburgh taught me a lot about anchoring decisions in facts. As a former prosecutor, he was adept at prosecuting the facts and the fact-bringers to develop the correct decision path.

During our debates in Japan, I often thought to myself, *How tragic that some thirty-two years after Three Mile Island, we are again debating the possibility of a hydrogen explosion. Have we not learned anything in thirty-two years?*

There are several things to learn from our evacuation delibera-
tions. First, the communication process could have been better.
Second, it appears that Washington created a crisis of its own that
should never have occurred. Governor Thornburgh calls this phe-
nomenon "emergency machoism." Some leaders just want to be
"seen" as a part of the crisis response. They cause chaos. I prefer to
believe that the friction between the U.S. and Japanese governments ultimately had no real impact on the Japanese people.

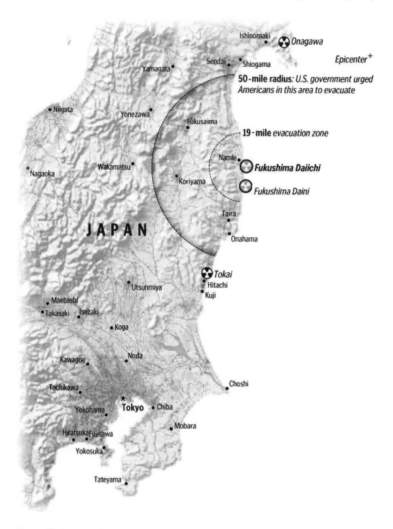

Figure 7. Evacuation zone map

On one of my later trips to Fukushima Prefecture, I determined that the locals had paid little attention to the U.S. evacuation advisories in any case. Had the Japanese agreed with the Americans on a fifty-mile evacuation zone, big cities such as Koriyama and Fukushima would have been evacuated. Where would the people have gone? Is it possible that the Americans didn't even look at a map when they came up with the fifty-mile recommendation?

Another problem involved the NRC chairman. When he went to Congress to testify, his mission should have been to assure the American people that we were on top of the situation, and that everyone in the States was safe—but that wasn't the case. His focus was misguided. Extreme-crisis leaders must understand that political and other unrelated considerations will always complicate an already dire situation.

CHAPTER 12

FEARS FOR UNIT 4

"What we believe at this time is that there has been a hydrogen explosion in this unit due to an uncovering of the fuel in the fuel pool. We believe that secondary containment has been destroyed and there is no water in the spent-fuel pool, and we believe that radiation levels are extremely high, which could possibly impact the ability to take corrective measures."

—NRC CHAIRMAN GREGORY JACZKO TO CONGRESS, MARCH 16, 2011

Gregory Jaczko, then the chairman of the NRC, testified before Congress that the NRC believed the spent-fuel pool for the Unit 4 reactor was empty and dry, and that there was a potential for a large, uncontrollable radioactive release into the atmosphere. Some thought this could trigger the necessity to evacuate Tokyo.

The spent-fuel pools at Units 1 through 4—which contained a significant amount of radioactive material—were located on the top floor of the reactor buildings. There were layers of sheet metal meant to shield them from the environment, but the concern was that if the water drained out or boiled off, the uncovered fuel assemblies would heat up. That heat would cause melting of zirconium, zirconium-water reactions, and other adverse repercussions. There was the theoretical potential that the zirconium itself could burn in a very intense fire, which would serve as an energy source to carry the radioactivity from the fuel into the atmosphere. In addition, zirconium fires are exceedingly difficult to extinguish. There is always concern regarding the condition of spent-fuel pools, but there was more concern here because the Unit 4 pool at Daiichi held hotter fuel.

On March 15, we witnessed the worst-case scenario when the

Unit 4 reactor building exploded, just as Units 1 and 3 had done. But in this case, we couldn't discern the source of the hydrogen. There was a lot of confusion about why the building would explode, given that there was no fuel in the reactor, as there had been in the other plants. What had caused the requisite hydrogen buildup? It was very confusing at the time, because we lacked critical information.

About a month later, NISA and TEPCO representatives briefed us about a ventilation pipe that connected Units 3 and 4. It became clear that hydrogen from Unit 3 had traveled through this pipe into the reactor building of Unit 4, causing the blast.

The Unit 4 explosion caused debris to rain down on the spent-fuel pool, and there was a genuine fear that the stainless-steel pool liner might fail or be punctured, causing the water to drain out. Additionally, because the reactor was in the midst of a refueling outage, there was concern about the condition of the pool gates that allowed flow between the reactor and the pool. It wasn't clear if the protective gates were installed in the spent-fuel pool or not or whether the gates might fail. There is a backup design feature to prevent the water from draining out completely in case of a gate failure, but there would still be a reduction of water in the pool, and that was of serious concern. Adding makeup water to the pool is a manual procedure, and because there was no access to the building itself, no makeup water source was available.

In the first couple of days after the Unit 4 explosion, there was steam vapor coming from the spent-fuel pool—then, suddenly, it stopped. For several days, it looked as if there was no water vapor coming from the building at all. Had the heat evaporated all the water, or had weather conditions changed so we couldn't see the vapor? Spraying water into the pool caused it to start steaming again, indicating to us that the water had hit hot metal. Thermal images seemed to show several "hotspots" in the spent-fuel pool, although it was unclear precisely what these pictures meant.

To make matters worse, we had received word that the radiation levels outside the building were very high. As the bulldozers cleared debris, these levels went down, indicating to us that the dirt they were moving around was covering up some highly radioactive material. We determined that its source must be the fuel itself.

The list of issues went on. Given the location of the spent-fuel pools, we were concerned that any earthquakes and tsunamis that might follow would topple the pool itself. If there were a fire in the pool, the evacuation zone around the plant would have to be expanded significantly to say the least, because the energy of the fire would release large amounts of radioactivity. The condition of the spent-fuel pool was preoccupying our thoughts. After examining what data we had, we determined that TEPCO should add water to the spent-fuel pools as quickly as possible to offset any leakage or evaporation. Over the course of several days, our considered approach remained that the operators should inject water in the spent-fuel pools.

During discussions with TEPCO, one engineer informally told one of our engineers that they, too, were concerned with the water level in the spent-fuel pool and speculated that it might be empty. I received a request from the Japanese government to go to the emergency center at the prime minister's office to view a videotape of a helicopter flight over the spent-fuel pools. In the very-poor-quality video, taken during a high-speed flyby, just a few frames showed the debris on top of the spent-fuel pool. The Japanese tried to persuade me that a reflection off the top of the water in the pool was visible, but as hard as I tried, I just didn't see it. Neither did two team members who accompanied me. The Japanese, however, chose to see water, and continued downplaying the problem, spurning our advice, and only reluctantly accepting help when their own efforts proved futile.

Behind the scenes and without our knowledge, Kan and his government were concerned about the Unit 4 spent-fuel pool as well. After the accident was over, I learned that, in a discussion with Goshi Hosono, special assistant to the prime minister, Plant Superintendent Masuo Yoshida said, "We'll die if that pool explodes." There were two-and-a-half times the number of fuel assemblies in that spent-fuel pool than were needed to power the reactor. It had the highest heat content of all the spent-fuel pools. One estimate indicates that the Unit 4 pool contained ten times more radioactive cesium-137 than that released in the Chernobyl disaster—5,000 times more than was released by the Hiroshima bomb.

The pool was located on the fourth floor of a building that had experienced a huge explosion; its reinforced concrete walls had bulged by more than three centimeters. A helicopter picture showed that the blast had torn off the roof, exposing the pool to the atmosphere. Some thought that the building could survive another earthquake and tsunami, despite the damage. But if those events caused a crack in the pool, all the water could drain out, making it impossible to cool the fuel. This would increase the amount of cesium released into the atmosphere during the accident by *ten times.*

The U.S. government was concerned that the Japanese were not being forthcoming and were marching toward an irretrievably catastrophic outcome. This presented us with an enormous dilemma. We perceived eight or nine symptoms of a problem with the water level in the spent-fuel pool. Offsetting that was one grainy videotape that may or may not have indicated the presence of water (probably not, in our estimation). In the name of public health and safety, we firmly believed that injection of water should commence. We communicated this to NRC headquarters.

Five days into the accident, on March 16, 2011, NRC chairman Jaczko in Washington told the Energy and Commerce Committee of the U.S. House of Representatives, "We believe that secondary containment has been destroyed and there is no water in the spent-fuel pool, and we believe that radiation levels are extremely high, which could impact the ability to take corrective measures." His statement created a good deal of uneasiness and fear among many Americans and Japanese.

In that Congressional briefing, the NRC chairman made two mistakes. First, he lost focus of his role as chairman. It would have been better if he had limited what he said about the conditions in Japan, confining his remarks to, "We're helping them." His number-one role was to explain any danger that might exist for Americans. The chairman's second mistake was in flouting the age-old leadership adage, *Do not say more than you know.* Jazcko might have said that the staff had *suggested* or *informed* him that the pool could be empty. It was not his job to make predictions or assess conditions that he didn't really understand. How could he, when we ourselves did not?

I was stunned when I heard his statement because, in my recollection, I hadn't even had a conversation about the condition of the spent-fuel pool with him. I could only conclude that one of our team members had talked directly to the chairman.

The next day, we gathered in the office of Minister of Defense Toshimi Kitazawa to discuss specific actions for cooling the spent-fuel pools. Before this meeting, Ambassador Roos and I, along with others, had developed talking points. The gist was that the Japanese should come up with a sustainable solution for water injection—emphasis on *sustainable*. Water injection methods should be safe from any future earthquakes, aftershocks, and tsunamis. We brought some thermal images that illustrated our concern about temperatures in the spent-fuel pools. We were hoping that these would encourage him to take strong action quickly—anything short of "heroic acts."

Unbeknownst to us, the Japanese were already planning to conduct a helicopter water drop on the Unit 4 spent-fuel pool. Ambassador Roos and I were sitting on a couch in Minister Kitazawa's office when the doors burst open behind us and two men ran into the room. They handed Minister Kitazawa a cell phone and whispered to him something that we couldn't hear. He went to the corner of the room with the phone. The rest of us stood to leave, to give him some privacy.

When we turned around, we saw on a big-screen television behind us a live video feed of the helicopter water drop being broadcast on international news. The phone call was apparently from Prime Minister Kan, congratulating Kitazawa on this successful action. I think it stirred up some national pride, much like the actions of the Chernobyl helicopter pilots. On the video, however, we could see the wind blowing the water away from the reactor building. It's likely that almost none of it dropped into the spent-fuel pool at all. But whether or not it was helping seemed beside the point; the Japanese viewed it as long-awaited visual (optical) evidence that the government was acting to stop the release of radiation.

The action was a successful "optical solution." There would be many more of those over the ensuing months. After the helicopter water drops, the Japanese started to use a truck with a boom to pump water into the fuel pools. This way, they could measure how much water they pumped, but they still did not have any visual indication of pool level.

We continued to gather information about the condition of the spent-fuel pools, but it was extremely difficult to gather the data we needed. The Japanese used makeshift cameras and temperature-measuring devices on booms, helicopters, drones, or whatever relatively safe means they could devise. The evidence was that water temperatures were normal, but we wanted to understand how they were *interpreting* the information they gathered.

Using our own technical analysis, we concluded that there was at least *some* water in the spent-fuel pool. I waffled on this a little after seeing new photos of the debris in the spent-fuel pool. It was inconceivable to me that not a single bit of debris had fallen to the bottom and punctured the protective stainless-steel liner.

I remembered that some years earlier, at a similar nuclear plant in the United States, a small piece of steel had fallen into the pool and had immediately caused a leak. Given that, I believed that if a large metal beam had fallen into the pool, there would very likely be major damage and leakage. In the following weeks, images taken using a camera mounted on the makeup water boom confirmed a lack of adequate water supply to keep up with the evaporation. Even the Japanese admitted that the water level was desperately low and needed makeup water.

We had to inform the NRC chairman of this conclusion. The official transcripts record my uncomfortable discussion with him, as I explained that his earlier statement to Congress had probably been inaccurate. While the fuel pool was not dry, it certainly had a significantly reduced inventory and needed continuous injections of water to maintain a safe level.

The condition of the spent-fuel pools would remain a highly controversial matter for weeks (and years) to come. In the immediate aftermath of the accident, we were under pressure to make timely decisions, and we concluded that the spent-fuel pools, especially that of Unit 4, needed water as soon as possible. We were not going to retreat from that point of view based on a fleeting shadow of what *might* have been water on a grainy videotape. Quite simply, the consequences of our actions if we were wrong would be cataclysmic, so we took the more conservative route, allowing for the possibility of damage in the spent-fuel pool and doing our best to ameliorate a worst-case scenario.

Years later, one of our senior analysts would suggest that if we had not focused on Unit 4's pool, the Japanese might have missed the need to add adequate water to it, which would have led to full evaporation and fuel damage.

Several days after my conversation with the NRC chairman, I traveled with a couple of team members to talk with Defense Minister Kitazawa about the spent-fuel pools. We sat across a conference table from the Joint Chiefs of Staff of the Self-Defense Forces. Kitazawa was at the head of the table.

After introductions, we began to talk about the details. Even in the middle of this catastrophe, I paused to contemplate the significance of being in the room. I kept thinking, *How many Americans have ever sat in such a high-level meeting with another government, briefing and debating with foreign chiefs of staff and the minister of defense?* Also running through my mind was what a unique honor it was for me—a country boy from West Virginia who had never even visited Japan—to be in this important position. It was surreal and, I must admit, I was a little distracted.

Besides that, I understood that we needed to elicit technical information from them, so I turned the meeting over to my assistant team leader.

As I listened to the ensuing conversation, I thought about the fact that during World War II, the predecessors of these men had sat around a conference table much like this one in the wake of Hiroshima and Nagasaki, making life-or-death decisions after a nuclear event. I now felt I was a part of that overwhelming cycle. This was very profound for me. It struck me that we had gone from the Manhattan Project to the Fukushima Project, and I was now a part of that continuum. Surely it could be said that no other nation has been more affected by nuclear science than Japan.

Extreme-crisis leaders go by the best information they have, and when they find out they were wrong about something, they revise their plans. When we were faced with compelling information that supported our judgment—and almost no information contradicting it—we prioritized what we felt would best protect people. Even if our assessment had led us to take some unnecessary measures, these would have minimal negative impact.

We couldn't help being distracted to some degree by the Japanese government, but in the end, our pressure to take aggressive action played a role in getting the Japanese to act aggressively when action was sorely needed. The helicopter drops served the purpose of giving people hope, even if they did not ultimately solve the impending problems.

I have no second guessing about our course of action in Fukushima, because I know we followed the course we believed would protect the most people most effectively.

CHAPTER 13

KEEP CALM AND GET ORGANIZED

*"At some point during discussions with Ambassador
John Roos, [a senior NRC engineer] said something to
the effect that, 'If this gets much worse or if that
containment breaches, then I'm leaving.
I'm going to leave the country.'"*

At the Nuclear Regulatory Commission, the people in headquarters and the first two responders to Japan, the senior reactor analysts (SRAs), were essential to the American response. Literally hundreds, if not thousands, of NRC staff worked exceptionally hard in those first days and throughout the response and recovery to the accident. These are extremely dedicated and talented people. In particular, the two SRAs who responded to Japan in those first few days contributed significantly to the American success in the embassy, along with the Institute of Nuclear Power Operations team in Atlanta and those INPO people dispatched to Japan, as well as the consortium of nuclear industry people in America. They are all remarkable individuals. I hesitate to list names, because I'll leave someone out, but each person who worked on the response gave his or her best. Those Americans can be extremely proud of their work.

After addressing some of the immediate crisis and safety issues, I could finally turn to organizing our efforts. In that first week, more than ever, I was *thinking about thinking*, or meta-thinking. Epiphanies can help you construct a mental model of the future while you continue to see what is immediately ahead. When facing a crisis, it's important to hit the pause button long enough to *think about thinking*. As a leader, you have to foresee the decisions and place your staff where they can bring you the proper facts when decisions need to be made.

We knew we had considerably more incorrect than correct data. We needed to navigate through random, fused, and missing data to organize the groundwork—to establish a battle rhythm, interpret the technical conditions, and provide an interpretation to the ambassador and other politicians that would help them grasp the picture.

In an extreme crisis, under unfathomable conditions, it is especially tricky to make sense of the situation you face. Developing a mental model in a complex, volatile situation is very difficult. High-risk-oriented organizations, such as the military and first responders, use an *incident command structure* (ICS), in which different entities within the organization are responsible for different elements of the response. After Fukushima, TEPCO implemented ICS at their Kashiwazaki-Kariwa Nuclear Power Plant.

At Daiichi, I think the challenge for Plant Superintendent Yoshida was that all the elements of an incident command fell on his shoulders. He was the most qualified, most trained, most knowledgeable, and most experienced person on-site, but it was still an unbelievably onerous burden for one person to try to organize and lead the people through those circumstances. At Daini, Masuda was spared having to deal with exploding buildings and high radiation levels; nevertheless, he faced a monumental challenge as well.

I knew from previous experience that one of the key elements of establishing a mental model is mastery of the technology. You must have a thorough and in-depth understanding of the technologies available—how they work and what their limitations are. In retrospect, the thermal imagery greatly misled us. We did not have experience with it and didn't understand it, but we relied on it heavily to develop our mental model.

As a side note, I sometimes wonder if that isn't what happened during the Iraq War with the storage of weapons of mass destruction. Did we have complete mastery of the technologies? Were our information-gathering systems reliable? Did these issues lead to poor decision-making? I think that in Iraq, possibly, as in Fukushima, we just didn't have sufficient mastery of the technology available.

Governor Thornburgh told me that at Three Mile Island they had recorded a very high radiation reading, much as we did around Fukushima Daiichi. In that case, the reading came from a helicopter. They nearly evacuated hundreds of thousands of people based upon that reading—but luckily, Thornburgh was cautious enough to reexamine those facts and interrogate their source. He found out that the helicopter had flown over the plant's exhaust stack, which emits concentrated radiation that then dissipates as it travels through the atmosphere. Only genuine mastery of the technology can lead to a precise interpretation of the facts.

Because of the extensive damage at Fukushima's nuclear reactor sites, we lacked suitable technology to formulate an accurate model. At one point, we tried to understand the structural integrity and stability of the reactor buildings after the explosions. To calculate the risk of the Unit 4 spent-fuel pool collapsing, we put out a request to the U.S. government and military for somebody who could give us a bomb damage assessment. In response, the military laughed at us, saying, essentially: "We don't try to keep a building standing. Our goal is to knock the building down, then we look for thermal images inside. If there are still thermal images, we hit it again. We aren't the ones to help you."

Before my arrival, some of the NRC engineers were looking at thermal images of the reactors and spent-fuel pools to diagnose the condition of the reactors, despite their unfamiliarity with this technology. I didn't immediately recognize this lack of mastery; it was several days later (around Thursday, March 17) that I realized we had no idea what these images were telling us.

The NRC engineers had been briefing Ambassador Roos for three days before my arrival, and had spoken to TEPCO and NISA. These people had looked at the thermal images. I made the critical mistake of accepting their conclusions without fully understanding the technology and their mastery of it. Quite simply, I deferred to expertise that wasn't there.

These NRC engineers had developed an easy working relationship with the ambassador, often going into his office or the embassy residence to put up their feet. While there, they held casual conversations and engaged in a lot of speculation in front of the ambassador, who wasn't trained to understand and deal with that information.

For instance, they would look at a thermal image and muse aloud, "*Well, it could be this, it could be that, or it could be the other thing,*" or they would get a piece of information from other sources— maybe news media or other human intelligence on the ground— and reported what the data *might* mean. To be blunt, they were whipsawing the ambassador.

Later, when we had the opportunity to go up to Fukushima, I found out that one of the senior NRC engineers had a phobia about radiation. At some point during discussions with Ambassador Roos, he said something to the effect that, "If this gets much worse or if that containment breaches, then I'm leaving. I'm going to leave the country." This fear is precisely what you *don't* want your nuclear expert to express to the ambassador in the middle of a nuclear crisis.

The NRC engineers' speculation inflamed the situation and panicked top American leaders. Adding to the confusion, a Japanese NISA director said informally to one of the NRC engineers that he believed there was significant core damage. The NRC engineer conveyed this conversation to the ambassador, and the lapse in proper protocol created major chaos. I think it was probably the impetus for the ambassador's infamous phone call to the Japanese prime minister in which he suggested that our people should go to their emergency center and watch from there. As a result of this suggestion and other things, the Japanese concluded that the ambassador wanted to take over the response. I feel quite certain, without direct evidence, that somehow the prime minister found out that the NISA director had posited core damage, and asked him to retract his statement (even though there was, in fact, core damage).

Maybe "panic" is too strong a word for what was happening, but at the very least the briefings and even the banter from NRC engineers sowed disorder. The leadership lesson to be gleaned here is about *fit*: When you send people into an event, you must make sure they are the right personnel for the situation. Think about that before you insert even your most technically trained people. Consider sending a manager first, or at the same time, who has the right personal qualities and diplomatic skills. In many situations, deference to expertise (without considering other factors) can lead you astray.

ESTABLISHING BASELINES

I continued to request that NRC headquarters provide us with a thermal image of what a spent-fuel pool *should* look like, from a similar reactor in the United States, so that I could get a sense of the heat signature of a normal, functioning spent-fuel pool. I can't say that they ever grasped the necessity, much less the urgency, of that request, and that was a major lapse. If I had known what a baseline thermal image looked like, I might have assessed the information I was receiving differently. I could have ameliorated the distracting panic growing around me with actual facts. At some point, I started losing confidence in the thermal images and essentially ordered the analysts to put them aside and join the rest of the team. To my mind, that information had become useless and potentially harmful.

It cannot be emphasized enough what a significant crisis-leadership lesson we can learn from these technological lapses. At some point, information technology may fail you—either because it isn't functioning properly or because the people in charge of interpreting the data do not have complete mastery of the technology. Leadership is about applying wisdom, not technology. Leaders must use their own wisdom and experience to make sense of a situation, thinking through it three steps ahead of their staff and remaining both flexible and in control.

Because our assumptions were changing minute by minute, the analysis of the extent of core damage varied widely. Earlier, I talked about the doubtful efficacy of the radiation-plume models. Modeling might be a good predictor of some events, but it does not provide a good basis for decision-making during an extreme crisis.

As a crisis leader, you must understand how and where you are getting information, how to process it, and how to make sense of the data. A colleague at NRC once described the lack of data as "swimming in an ocean of silence." The NRC chairman commented that we were working in "the fog of war." We were getting some information—not a lot, by any means—from the Japan Atomic Industrial Forum, and it was flowing back to Washington without discipline. Raw data was being shared along informal channels in every direction. This made it extremely difficult to identify or process accurate data, and confusion reigned.

As in any extreme crisis, we were trying to analyze, think, and respond at the same time. There were many challenges and no feasible way to address them all. At one point, I remember thinking, *There were only two to three big issues to begin with, but if we ask TEPCO and NISA what their priorities are, we'd get a different answer every day.* This was partially due to all the changes occurring at the plants, but mostly because of conflicting data sources.

Backchannels of communication developed among people who had long relationships with the Department of Energy. Those channels encompassed the White House, which had almost daily phone calls with academics and others within Japan who, it seemed, didn't accurately grasp the situation. This became very challenging for us because the White House was getting a completely different picture from what we were experiencing on the ground. This damaged our credibility, as Washington knew the ambassador was getting his information primarily from us, while they were getting it from Dr. Shunsuke Kondo, "the father of nuclear power in Japan," and others. Some in Washington had long-term relationships with Dr. Kondo. What they didn't know was that he himself would later tell me that based on what happened at Daiichi, he would have received a grade of "F" for his participation.

I had to assess and reassess how our team worked in very short order. Besides structuring our internal team roles and responsibilities, I needed to structure our approach with the ambassador and all those with whom we were communicating.

LEADERSHIP STYLE

On that first flight to Japan, I began thinking about my leadership style. I recalled one of my great role models, Harold Denton, the NRC person who had served as the on-site executive at Three Mile Island. When considerable confusion arose between the White House and NRC headquarters, President Jimmy Carter had asked the NRC to designate someone on the ground to be answerable directly to him and Governor Richard Thornburgh, thereby cutting out the NRC Washington headquarters. That role fell to Denton. That situation came to mind when the 2011 White House asked the NRC to dispatch someone to Japan.

I knew that when Harold—a calm Southerner with a slow drawl—arrived at Three Mile Island, he brought much-needed calm to the situation. Sadly, Harold passed away during the writing of this book, but the takeaway from my conversations with him is that he was a good leader—humble, cheerful, comforting, thoughtful, and reasonable. I owe him my gratitude for having taught me many valuable lessons that came into play during the Fukushima accident.

Understanding that Japan's business culture is rather formal, I didn't copy Harold's casual style, but I did honor my country roots by remaining plainspoken and making liberal use of colloquialisms. Like Harold, I tried to remain calm and collected and add a reasonable voice to the mix. His wisdom was on my mind when I set about controlling the information flow to Ambassador Roos.

I insisted that all information provided to the ambassador come from me, so that he would be getting a clear, accurate picture of the situation and straightforward recommendations. Ambassador Roos seemed comfortable with that arrangement, which kept him out of any back-and-forth debate that might muddy the waters. The debate was to come before the briefing.

It appeared that pandemonium was rampant in Washington. I noticed that the less data there was, the more confusion, and came to call this the *Casto Pandemonium Curve*. I had to find ways to inject coherence despite the lack of reliable data, and to describe technical situations in a way that nonprofessionals could understand. I became a situational interpreter. For instance, at one point I told

Ambassador Roos that sheltering people in their houses for a nuclear accident is like having people stay inside during a snowstorm, except without the snow. At another point, I shared with him that we were working toward cooling the reactors and spent-fuel pools so that any radioactive releases could be measured in meters, not kilometers. Those kinds of explanations provided the requisite clarity for the boss.

The world was watching the Fukushima Daiichi accident. As participants in the first live, web-streamed nuclear event, it was up to us to quickly interpret highly technical information for politicians. To improve communications with our Japanese colleagues, we had to speak their language and show that we understood the situation. By *language,* I don't mean Japanese, but technical language; we made a concerted effort to use their terms. A small example is the terminology for the physical reactors—we would normally have called the site *Fukushima.* They called it "1F," and we quickly learned to do the same. This kind of simple adjustment led to better clarity between our teams.

At times, internal communications within the embassy were a challenge because random talk and overheard comments could be taken out of context. Hallway talk travels far, and it was important to structure strict lines of communication within the confines of the management team.

Further problems developed in our attempts to get accurate, timely information from TEPCO and NISA. The problem was that no individual TEPCO engineer had all the information, so we were always piecing together disparate accounts and couldn't construct a clear picture of what was going on. The goal of Hisanori Nei, our NISA counterpart, seemed to be to limit the amount of information we received. I'm not sure why he wanted that—whether he was acting on his own or on orders from others, or whether it was just his nature. Mr. Nei posed a challenge throughout the accident, but he was the only source I had within NISA. Sometimes, you must work with the people you have and manage the personality dynamics.

As technical people, we tend to prefer to get the facts from the people closest to the source of information—ideally, engineer to engineer. But, in the case of Fukushima, the fundamental communications flaw was that we *only* had engineers talking to engineers: NRC engineers talking to TEPCO and NISA engineers. Leaders must always strive to get the big picture, to see everything with the widest possible lens. As a leader, you must encourage your technical leadership to talk to technical leadership (including at political and government levels) and related experts, to make sure you are not getting isolated pictures of the situation. It's easy to overlook this important issue in the detail-deluge that occurs during a crisis.

In my case, perhaps I didn't get to Mr. Nei soon enough to establish a better working relationship with him. Because I was spending the bulk of my time with the ambassador, I had little time to get out and gather data myself. I soon realized that I needed to break down whatever barrier there was with NISA and get acquainted with those at the highest level. It took me a couple days before I established that relationship.

I traveled to NISA to apologize in general, and specifically to Mr. Nei, for not talking to him sooner. I explained that I had been busy with the ambassador and thanked him for the information they were providing. I apologized to him in front of my staff so that he would see that I was humbled and acknowledged having made a mistake, then apologized to him again privately. During that conversation, I found out that he had lived for a long time in Houston. Since I'd lived in Dallas for a few years, we talked a bit about Texas. We began to develop a relationship that has lasted for years beyond the accident.

When stepping into a crisis, it's important to survey all the players, prioritize them in terms of their importance in addressing the crisis, introduce yourself to the top dog in the country or facility, and establish a direct line of communication with that person.

Next, I had to find a method to convey the "direction" of the accident—whether, at any given point, it was getting worse or better. I structured my reports to the ambassador in terms of goalposts, which I framed as the *best-worst case* and the *worst-worst case.* (There were no "best cases," unfortunately.) I established the ranges within the goalposts, so there was no need to convey information explicitly. I simply had to tell him that we were still between the goalposts. If I received what I considered to be reliable information that indicated movement outside the goalposts, I shared it with him; otherwise, I just let it "bake."

Setting goalposts and not chasing information and data were very important to navigating Fukushima, as they always are in a crisis. Chasing the data can make you second-guess or change your strategies unnecessarily. When we saw that there was water going into the cores and no radiation coming from the plants, we deemed our strategy successful—and we did *not* want to change direction. If data came in slightly outside the goalposts but not drastically so, I let it bake and monitored it to see if that data point collected "friends," or more data around it. I continued my strategy of not sharing every scrap of information with the ambassador until it was thoroughly baked, thus preventing a repeat of what had happened at the beginning, when the reactor analysts had whipsawed the ambassador with unfounded information.

REPORTING TO THE AMBASSADOR

At some point after the Unit 4 explosion, I went with my gut and decided that the worst had already happened. I remember making a small joke to the ambassador, that he could sleep for twenty minutes that night because things were stabilizing; we had a chuckle about that. Each day, depending on whether the news was good or bad, I'd say, "Well, now you can sleep for *twenty-two* minutes," or "*eighteen* minutes"—you get the idea. I felt injecting a little jocularity relieved some of the tension, and at the same time, I was using this simplistic joke to clarify where things stood.

Each time I went up to the ninth floor of the embassy to brief Ambassador Roos, I put on my jacket to convey that the visit was official. I told him only what I knew, in the style of a classic executive briefing. It was not until after we'd developed a rapport that I became somewhat more casual with him, but never in front of his staff or mine. If he and I stayed back after a staff meeting and he asked me directly what I thought, I would tell him honestly.

The ambassador seemed to have full confidence that I was monitoring and assessing the situation moment by moment. One day,

I sat with him alone after a meeting and said: "Ambassador Roos, I don't think this is going to get any worse. I think the worst has happened. I think you can relax. I'm going to tell you what I must tell you every day, but absent a major upset, I think you should know that I doubt it will get any worse." I'd made this judgment because there was no energy left—the hydrogen source was depleted. There was now no mechanism to send significant amounts of radioactivity into the atmosphere. I reported this to the ambassador only after I'd established a sound communications structure and a trusting relationship with him.

At that point, the ambassador and I worked up an outline for the kind and depth of information he wanted. We came up with a "Windows Chart" that listed the status of the reactors, spent-fuel pools, and reactor-building containment. At first, I provided more of an overview, but as the months progressed, my reporting became more precise and accurate. From March to December, as conditions evolved, we altered our chart to be most relevant and, of course, to reset those goalposts, making the chart one of our most valuable tools in managing the crisis. As time moved on, the ambassador focused on containment parameters. Roos and James Zumwalt (the embassy's deputy chief of mission) had acquired enough knowledge at this point to understand that they needed to monitor containment more than the reactors themselves.

The fact that there wasn't a door on the conference room didn't help my attempts to rein in the flow of information. Embassy staff could stroll in at any time and look at our damage pictures and what we had written on the board, attend our briefings, or overhear us speculating. Unfortunately, we didn't have what I refer to as the *horsepower* to exclude other federal agencies, such as the State Department. I am quite certain these casual observers carried tidbits back to their families, to other embassy workers, and to Washington. Some of them even took notes at our team meetings, which they sent to Washington before I could share the information with the ambassador or NRC headquarters. Some of this was data that I intended to let "bake," or that I was still assessing for accuracy, but off it went, through uncontrolled channels, fostering a great deal of confusion and, in some instances, mayhem. These interlopers were often not technically competent enough to possess an accurate perspective. At one point, the State Department brought in a person from elsewhere in Asia, specifically to sit in our room and listen. She told me of her role, about which she had the good grace to be embarrassed.

Behind the scenes, I urged my staff to confine themselves to the big items we needed to cover but hold back any raw or preliminary information that still called for analysis. Then I started conducting a meeting *after* the meeting to discuss certain things among ourselves. We did dial in the external nuclear experts, such as the team from the Institute of Nuclear Power Operations, and were grateful for their help. I wanted them present during those behind-the-scenes briefings, because INPO was a big player, a reliable resource for us, and its team understood the situation on a level that made their ideas valuable.

Sometimes the embassy staff wandered in and out of the conference room carrying partially understood bits of data and rumors to the other staff or dependents in the embassy compound. Those conversations generated unnecessary angst. To allay and counteract it, I had members of my team travel to the embassy compound often, to share accurate information on the status of Fukushima.

CHAPTER 14

"HELP ME"

We got the stiff arm and were sent away with,
"We don't need your help right now. We'll call you when
we need you." (And, for good measure, they further
disparaged their chairman.)

We were notified on March 18, 2011, that TEPCO Chairman Tsunehisa Katsumata wanted to talk with us. I don't remember who arranged the meeting, but I think the Japanese government was in favor of it because TEPCO had a message for our government. Ambassador Roos and I were to travel to their headquarters to meet with the assistant to the prime minister for the disaster, Goshi Hosono, and had very little time to prepare for the meeting.

We'd had several meetings with TEPCO and government officials prior to this, mostly when the potential need for "heroic acts" was being discussed, along with rumors that TEPCO would abandon the site. I assumed that this meeting would be a continuation of such talks.

At the back of TEPCO Headquarters, in an underground parking area, we were met and ushered up a private elevator to the top-echelon offices. Stealth was required so we could avoid some ten thousand protesters who had stationed themselves permanently in front of the building. We sat down with Chairman Katsumata and TEPCO Vice President and Chief Nuclear Officer Sakae Muto in a sparse, overheated, and dim conference room. I certainly had no need for the warm tea brought by an assistant.

The day before, the TEPCO chairman had held a press conference during which he broke down and was ushered—carried, really—from the room. I knew that this gentleman was devastated by the events and offered the sympathy of the U.S. government, our condolences for the lost citizens of Japan, and our desire to help in every way we could. Muto discussed the conditions at the plant as he understood them at the time, mainly talking about the injection of seawater.

It hit me hard that these TEPCO officials were willing to admit just how grave the situation was. They seemed to feel that there was little chance of recovery from the event. Before this conversation, I had thought that TEPCO would be confident in its ability to limit the damage to the reactors; now I had major doubts. They did not seem to share my assessment that the worst was over—and if they did not persist in their efforts to cool the cores and spent-fuel pools, they might be right.

The chairman asked us to bring our personnel to TEPCO, where we would work directly with its emergency response manager. He repeated this request several times. As we were wrapping up the meeting, the chairman—a frail, white-haired, dapper man who appeared to be in his seventies—looked at me and said, "Help me." Those two words struck me harder than anything else I'd heard that day. Here was a wise and honorable man with many resources and talents, who had suffered an emotional breakdown the day before, asking—poignantly, in English— *"Help me."*

This sense of desperation naturally raised doubts in my mind about the man's fitness to lead his people through the disaster. At the same time though, I found myself respecting his dedication, his sense of professionalism, and his desire to do the right thing.

I knew how integral the sense of honor is to Japanese culture. If the chairman had swallowed his honor to ask an American for help—in English—this was a serious situation. Mitigating the disaster became a personal fight for me at that moment, and I committed myself to doing everything I could for this gentleman.

The chairman and Muto escorted us back to the private elevator, another ceremonial, cultural act that I found compelling. At the elevator, they bowed to us over and over until the doors closed. This accident was much bigger than I'd allowed myself to believe until that moment. *The chairman of TEPCO is telling us that they cannot handle this event. The utility and the Japanese government cannot deal with the accident themselves.* I could feel my blood pressure spiking and my adrenaline flowing. I immediately called Washington and told them that we needed more help. I was already working out in my head how many areas we would need to tackle—whether we needed off-site and/or on-site mitigation and what specific areas of expertise our people would need to have in order to help TEPCO.

Although I was cognizant that our rules of engagement necessitated government-to-government effort—that we were not mandated to work directly with the utility—I hoped that somehow this would resolve smoothly. As it unfolded, the constraints we were under caused us much difficulty throughout our response. While I fully grasped that having rules of engagement was necessary, I couldn't envision that our government would not extend help to TEPCO when they specifically requested it. I would need U.S. nuclear industry people here immediately to serve as intermediaries between TEPCO and us, and I asked for them continually.

I was confident that I could trust our nuclear industry to share information with us, because I believed that they and the NRC would have one common goal—safely resolving the problems with the reactors. TEPCO had offered conference space in their headquarters—a small conference room that could hold up to five people. I accepted the offer as a means of penetration into TEPCO, even though I didn't think that we—the NRC—would end up using the space much, if at all. Any access to that building would be beneficial. More important, I knew that a massive amount of help from every quarter was needed in the worst way.

I thought about our nuclear response teams, how big they might be and how we could organize what we call a *base team*. On my way to rally our experts and staff and take them back over to TEPCO, I called Washington.

NRC Chairman Jaczko and some other staff got on the line and listened as I transmitted my thoughtfully compiled list of assets and personnel we needed. His reaction was not positive. The number of staff requested struck Jaczko as beyond our mandate to provide consultation and advice. Washington simply did not accept my assessment. At one point in the conversation, the chairman said to me, "Stop and take thirty seconds to take a breath." That struck me as demeaning. It wasn't as if I were panicking; I was using my decades of training, experience, and the expertise for which I'd been hired to *lead*. I understood the situation and its gravity. I knew what needed to be done and grasped the scale of what it would take.

The result of the call was clear: I would have to compromise. I determined to accept any resources they were willing to give, hoping we could revisit the situation later. If it meant that some of that team would have to be in Washington and some in Tokyo, I would make that work. But the fact that the chairman seemed to think I was overreacting got us off to an inauspicious start. I feared it weakened his confidence in me as a leader.

I did not tell the NRC chairman of the plaintive *help me* from the head of TEPCO. The people in Washington could have no sense of the realities I had been made aware of at that meeting, and could not at that moment appreciate just how dire the situation was. This was not to mention the rumor that TEPCO might abandon Daiichi; that the Self-Defense Forces seemed frozen in their tracks; that on-site TEPCO workers were paralyzed. This all seemed impossible to communicate at that moment, so I suppose it isn't surprising that Washington disregarded my requests. Clearly, the Three Mile Island accident had not proved sufficiently instructive.

Despite this discouraging response, we got underway as best we could. We gathered people from the embassy to go over and help TEPCO for a length of time no one could predict. But, by the time we got there, conditions seemed to have changed radically. Essentially, we got the stiff arm and were sent away with words to the effect that, *"We don't need your help right now. We'll call you when we need you."*

I was stunned that these people would defy their chairman's directives and send us away. When I reflected on his poignant plea for help, another interpretation occurred to me. The personal pronoun he'd used might have meant that he feared that his people were incapable of handling the situation and that they were failing him. Had he already measured their willingness to throw him under the bus? Perhaps that's why he had spoken in English; why he had whispered quietly, "Help me." It might have been a personal statement that he lacked confidence in his own staff and their ability to handle the disaster.

During this whole monumental event, filled as it was with unforgettable moments, this was among the most notable to me. I was moved on both a personal and professional level by my intimate glimpse of this man within this situation. Nevertheless, a criminal indictment against Chairman Katsumata and two of his vice presidents—including Muto—would be handed down in 2017, for their failure to act on data that had been presented to them that a tsunami higher than thirty-two feet (the height of the Daiichi seawall) could flood the plant and result in a power outage. Had the seawall been expanded, the tsunami may not have reached Daiichi.

As an extreme-crisis leader, one must be prepared to react to even subtle signs of helplessness; these are crucial road signs. In some cases, helplessness may be obvious, and in others, much less so. It's crucial to assess the capabilities of other leaders and their organizations. I will never forget the act of defiance on the part of TEPCO's staff against their own chairman. This was one of several acts of defiance during the accident. Some of the defiant acts, like continuing seawater injection or assuring the safety of operators in the field, were good, some bad, like rejecting Chairman Katsumata's request. During an extreme crisis, defiance can be a wicked challenge.

CHAPTER 15

FLOOD THE BUILDING

Listen, learn, help, and lead.

When everyone is pulling in the same direction, leadership is satisfying. But when people are pulling in many directions and some of them are diametrically opposed to your ideas, it tests the mettle of the leader. I call this leadership lesson, "Listen, learn, help, and lead." Often, in international situations, we Americans go into a country and listen to the locals discuss their problem, then analyze what they're telling us, and then boss them around because we think we know better than anyone else. There's no doubt this happened in the first week of the accident. I call this all-too-common leadership model, "Listen, analyze, and boss around," and consider it highly flawed, because all too often, we are not actually listening to our international counterparts but to ourselves.

As I have described, when reactors are deprived of essential cooling water, core damage occurs, as it did in reactors 1, 2, and 3 at Fukushima. We feared that the extreme heat that resulted might cause the bottoms of the reactor vessels to fail, allowing the molten fuel to pour into the basement of the containment building. Over the years following the accident, this situation was confirmed to at least some degree on all three of the reactors.

In the immediate aftermath of the accident, the Japanese used firetrucks in an attempt to flood the vessels and cover the fuel. Some water was going where it was intended, but, because of breaks in the buildings and vessels, much of it was leaking into the outer containment buildings. Then, because these structures were also damaged, that water was flowing from the containments into a secondary building housing the turbine generator. This flow path caused highly radioactive water to flood the bottom floors there, and could ultimately reach the ocean.

Of course, initially, those responsible for pumping had no understanding of these potential flow paths. The emergency procedure for reactors provides guidelines to the operators on the minimum amount of water needed to flow through to keep the cores cooled. The output from the firetrucks was below this minimum. Our own established emergency procedures would instruct the operators to flood the containment building itself, to cover the molten fuel on the floor. This action would send water up through the bottom end of the reactor vessel and—we hoped—to where any remaining nuclear fuel was, cooling that fuel as well. The water in containment would also serve as a shield from the highly radioactive fuel.

Our team was insistent that the Japanese flood the containment building, as required by our emergency procedures, and install a robust water supply. We shared with the Japanese this protocol, which had been devised by General Electric, the company that designed the reactor. Much to our dismay and confusion, the Japanese would not consider flooding the containment.

By Sunday, March 20, I had attended one of the technical meetings and listened to our staff talk to the Japanese government and TEPCO about flooding the containment. I had heard their rationale against flooding the containment, which was based on their fear that the building would collapse under the weight of the water. They said they had no way of telling how high the water level was inside the building, and I acknowledged that fact, but I felt there were measures they could take to ascertain that level. They saw this as an obstacle, but the biggest obstacle, as they saw it, was the cracks in the building structure itself. Those cracks meant that the turbine building basements were filling with water—and there was the potential that it could leak into the ocean. They were under orders from their prime minister not to release highly contaminated water, so their only recourse was to minimize the flow from the firetrucks to prevent overflow. There was the rub: The low flow rate they were producing wasn't going to be enough to cool the reactors—or what was left of them—but they believed it was guarding against contamination of the ocean.

The low flow also meant that the reactors would take much longer to stabilize. In the end, it took nearly six months for the reactors to achieve what we call *cold shutdown condition*. This was months longer than it should have taken, because a decision had been made based on politics rather than science. Perceived environmental concerns (which were inherently political) trumped the mechanics of how best to solve the problem.

As I said earlier, it is not unusual for political considerations to override technical decisions. It happened during Hurricane Sandy (unofficially referred to as Superstorm Sandy), which hit the east coast of the United States in October 2012. In that case, local state governors wanted to keep cars off the road, but it was a presidential primary election time and the federal government wanted to ensure that the people could exercise their right to vote. So, the polling stations were open, encouraging people to drive in conditions the governors believed to be unsafe.

In the case of Fukushima Daiichi, the Japanese government intervened four times, prioritizing their political considerations over the technical conclusions of experts. The government was wrong in every case. The conflict started with the first evacuation, when the government wanted a ten-mile radius rather than the five-mile radius that Yoshida had the right to implement. Expanding the evacuation zone delayed the venting process, contributing to the Unit 1 explosion. Then there was the matter of the injection of seawater, which the government opposed because it didn't want to ruin the reactors. After that, the government blocked the release of any radioactive water into the sea, which delayed the cold shutdown of the reactors by months. And, finally, there was the moment when Prime Minister Kan went to the Tokyo headquarters and, in a live-streamed speech, demoralized the operators, rather than support and encourage them as they labored to save their country from disaster. These government or external interactions (the Japanese called them "interferences") represent a factor that the extreme-crisis leader must acknowledge and attempt to mitigate.

As I listened to the Japanese argument against increased water flow, I thought: *I understand their position. Given the constraints they have, there's no option to flood the building.* I had told Ambassador Roos that I didn't believe there were any zero-release solutions, meaning that eventually they would have to let radioactive water flow into the sea. But they did avoid flooding the turbine building basement, devising a system that would recirculate the water from that building back to the reactor vessel. Operators installed filters in the recirculation path that filtered out the highly contaminated material, returning it in a much cleaner state to the reactor vessel. The limited water flow to the reactors did contribute to further core damage. To their credit, I think the Japanese did the right thing there and, happily, the process worked.

After that meeting, I told our team never to mention flooding the containments again, that the Japanese were correct, and we should not push them, despite our own emergency guidelines. This situation points up the way Americans often approach international partners. We often walk in, get some facts, analyze the situation, and come up with what we believe to be the one and only correct answer. We then issue an order. "Flood it," we insisted, as if it were not our opinion but the only "proper" way to proceed. But that rigidity threatened to cause a grievous communications gap with the Japanese, one that we could only repair by backing off.

At my insistence, we would no longer follow the American model of listening, analyzing, and bossing. In order to help the Japanese contend with their fractured buildings and prevent radiation from leaking into the ocean, we would slow down and listen to them fully, which is tremendously difficult in a crisis. Then we would work with them to create plans and solutions on a solid, realistic foundation that worked for them. We would "listen, learn, help, and lead." That is, listen from their perspective, learn the issues as they see them, help them solve the issues as they see them, and then perhaps with the trust that this process builds, lead the way to a good solution.

This protocol is applicable in a multitude of situations. It applies when dealing with children at home and colleagues at work. If you're not in a position of absolute authority, then you must persuade and, ideally, inspire people to move in the direction you see as best. See the situation through their eyes, help them through the difficulty as they see it. Only then can you lead effectively.

CHAPTER 16

RADIOPHOBIA

The anomalous reading created a bit of a culture clash,
something that we had to work through as leaders during
the first several weeks of the response.

The NRC, the U.S. State Department, and the Pacific Command attempted to address public fear of the situation by providing information and technical support for our communications with the public. Public relations problems were undermining our best efforts. Anti-nuclear activists and contrarian nuclear experts (who lacked detailed information about the unfolding crisis) took to social media to speculate about the condition of the plant. This "fake news" fueled the panic and undermined our best efforts to calm the fears of embassy dependents and the public.

One American anti-nuclear activist considered himself an expert on everything. He would take one tiny factual detail and fabricate a fairy tale from it. American embassy dependents watched his videos and believed his falsehoods, which then made their way around the embassy. This disinformation undermined our credibility with the embassy staff because they didn't know whom to believe.

After seeing a particularly inflammatory video, I entertained the thought of going rogue and contacting the creator and disseminator directly to ask him if he had any concept of how much damage he was causing with his wild speculation and uninformed opinions. I immediately thought better of it, but I could only conclude that radiophobia—fear of radiation—might be the one toxic thing that spreads faster than radiation itself.

During Fukushima, the moment came when enough Americans wanted to leave that the State Department decided to allow voluntary early departure. In advising on that decision, we tried to use good science as a basis; ultimately, however, emotions won out over rationality. Citizens did not feel they had control of the risks that threatened them thus they wanted to leave.

On March 21, 2011, the State Department announced Dependent Early Departure, meaning that dependents could return home to escape a dangerous condition. We debated this declaration for days because of the science of radiation and the extent of the immediate and long-term threat posed—not to mention the delicacy of the political aspect. Our departure could be perceived as Americans fleeing Japan. Other governments had already evacuated their Tokyo embassies to western Japan. Any evacuations by Americans would make it difficult for the Japanese government to reassure its own people that they were safe. Naturally, the Japanese people asked themselves, *Are our institutions incompetent or dishonest?*

The problem, in retrospect, was that we had rung a bell, as it were, with the departures, and couldn't unring it. After the proximate danger had passed, we had a difficult time luring Americans back to Japan, partially because we couldn't commit to the scientific reasoning that would convince them to come. If, as a leader, you use emotional intelligence to justify decisions rather than scientific data, it becomes difficult to find a scientific/technical reason to "un-justify" that decision. We're speaking here about evacuations, but the lesson applies to a host of other issues as well. The people were still afraid, the evacuation zone was still in force, the reactors were and would remain fragile for at least another six months—so how could we convincingly say it was safe to come back?

On March 14, a radioactive plume from Fukushima Daiichi had contaminated the USS *Ronald Reagan* and seventeen helicopter crewmembers. Other crews were exposed to low levels of radiation. As a precaution, the U.S. Navy prescribed potassium iodine pills to some crewmembers. On March 15, the U.S. Navy detected above-background radiation readings south of Tokyo, at the Yokosuka and Atsugi facilities.[1] Many vulnerable groups live on and near these facilities, so the readings rightfully concerned U.S. military leaders. Commanders were especially concerned about Department of Defense dependents living there, and directed base residents to remain indoors as a precaution.

On March 16, the State Department issued a travel warning: Americans were not to travel or remain within fifty miles of Fukushima or some additional "exclusion zones." In a travel advisory issued on the 17th, the State Department suggested that U.S. citizens not travel to Japan at all, and that those who were there should consider leaving. On March 21, the Department of Defense announced Operation Pacific Passage, which moved Department of Defense dependents (there were about 7,322) from Japan.

On that same day, the U.S. Navy detected inexplicably high radiation readings halfway between Fukushima Daiichi and Yokosuka Naval Base. The Naval Reactors Program (the NRC of the Navy) recommended that military and dependent personnel within a 200-mile radius of the plant be offered potassium iodine, and that Operation Pacific Passage be expanded to include those personnel. Commanders contemplated administering potassium iodine to dependents at the southern bases as well.

The NRC was skeptical about the accuracy of the reading, and in the end, it turned out to be random. What it revealed, though, were the differences in organizational cultures between the Naval Reactors Program people and the NRC. The Naval Reactors organization, rightfully so, is extremely cautious about any amount of radiation, because even traces released on a nuclear submarine or ship can turn into a crisis quickly; measures must be taken immediately to protect the lives of the sailors, who are not able to evacuate and might not even be able to call for help. The NRC, on the other hand, deals with big commercial reactors that routinely put out a higher level of radiation. They have more experience with small leaks.

The anomalous reading created a bit of a culture clash that, as leaders, we had to help resolve. Troy Mueller, director of the Naval Reactors Program, and I clashed many times. I remember one par-

ticularly intense argument about who would advise the ambassador
and what that advice would be. As it turned out, Troy and I became
close friends later, and still are.

If conditions at the reactors continued to get worse, it would be
prudent to evacuate the bases. As a precaution, the U.S. military in
Japan and the embassy began executing their Emergency Action
Plans.[2] These plans (prepared for all U.S. foreign bases and embassies) protect classified documents from unauthorized access. Huge
shredders at the U.S. embassy hummed for four days, eating classified documents. Burn bags at the bases were stuffed with documents and burned. Classified machines and servers were crushed.
This was all done on a sliding priority. Nonessential personnel destroyed their documents and machines before leaving on Operation
Pacific Passage. In the worse-case scenario, military leaders and
diplomatic staff would destroy their documents and depart as well.

While this kind of emergency contingency is common worldwide, Fukushima was a rare instance when such a plan was put into
action. Clearly, at least some American officials believed that U.S.
departure from Japan would become necessary and that it would
be prudent to take steps to protect American information and
sources. Simultaneously, a team of U.S. embassy personnel, understanding that commercial and military flights would be insufficient
to handle a mass evacuation, devised a plan that would evacuate
thousands of Americans via aircraft carriers. Two staff members
from each department in the embassy would remain, while the
remainder would be moved to a compound at Osaka, Japan.

Other nations were making similar plans to move their citizens
to safety. Japanese people who heard of these plans began calling
the foreigners *fly-jins*, a derivation of the Japanese word *gaijin*,
meaning foreigners.

Regardless of any safety plans, from March 11 on, conditions
within the country were unfathomable. Buildings were exploding,
radioactive plumes were being unleashed, and the natural disaster
continued with aftershocks and tsunami warnings. Fear was stoked
by bogus YouTube videos, and the widespread panic over this unprecedented nuclear accident led to moments of truth in the
American-Japanese relationship. It was understandable that the

American leaders would lean forward to protect its own citizens. The Japanese looked to their leaders for indications of the same kind of decisive action.

I believe that the anomalous reading of March 21 panicked some U.S. government leaders, who also seemed to wonder whether they were being deceived about conditions by the Japanese or even the NRC itself. Were we on the ground underplaying the accident—or, perhaps, underestimating it? Was Ambassador Roos perhaps a party to this "whitewashing" when he advised (on my input) that the worst was over? One thing is clear: The Navy reaction to the anomalous reading damaged confidence in the credibility of Naval Reactors, and they lost some influence at the embassy.

Random, invalidated data points can cause significant disruptions in the narrative. Lack of trust took hold between the NRC and the Navy teams, though I understood that the driver for the Navy's position was its low threshold for radioactive releases. Once I explained this, the NRC team gained credibility. Ultimately, Ambassador Roos tended to lean even more heavily on the advice from our team.

PROTECTIVE MEASURES

In the United States, during the "Plume Phase" of an accident, the NRC's Protective Measures team provides the governor with recommendations for measures such as evacuation and sheltering. The Department of Energy supports with data collection and analysis via aircraft measurements and field measurement teams. Once the event progresses to the "Ingestion Pathway" phase, when radioisotopes have settled on the countryside, Protective Measure recommendations come from the Department of Energy. At that point, rather than an inhalation hazard as in the Plume Phase, the radioisotopes pose a hazard from the consumption of contaminated foodstuff.

During the Fukushima Daiichi accident, the American model would face some challenges—the biggest being that there was no governor with powers to implement Protective Measures;

Ambassador Roos was the closest we had to such an official. All we could do was recommend actions to the Japanese and advise American citizens and military. There were other problems as well. The NRC Protective Measures teams, who were trained to provide recommendations, were not comfortable actually making decisions. We did have Department of Energy staff in the embassy gathering radiation data from many sources, including their aircraft, but they were not readily sharing the data with us and would offer no recommendations. The Department of Energy was holding its radiation information close because it wanted to provide that information to their counterparts in Washington and its own stateside labs for analysis. This situation was less than helpful to me, as the nuclear advisor to the ambassador. Further, the Naval Reactors Program had an extensive radiation monitoring system supporting their nuclear-powered ships—but we had no access to that information either. I felt extremely handicapped.

A month or so after the accident, the NRC recognized that the situation had progressed to the Ingestion Phase. Naturally, at that point, the Department of Energy would take over responsibility for Protective Measures. Accordingly, NRC officials asked the Department of Energy to find someone to take my place as the advisor to the ambassador. Before long, I was told that, after searching the Department of Energy, they could find *no* executive capable of replacing me in the position. *Really? Thousands of Department of Energy workers, and you can't find anyone?* This issue went to the White House Principles Committee, which promptly created an "action item" for the Department of Energy: Replace the NRC team leader. Well, that didn't happen.

• CHAPTER 17 •

KANTEI MEETINGS/ HOSONO PROCESS

THE ASAHI SHIMBUN

INTERVIEW/ YOICHI FUNABASHI: Fukushima nuclear crisis revealed Japan's governing defects

February 29, 2012

BY ROY K. AKAGAWA/ STAFF WRITER

In a word, between March 11 and March 17, the Japan-U.S. alliance was in a crisis. It appeared when the United States issued a travel advisory recommending not entering an 80-kilometer radius from the Fukushima No. 1 plant when the Japanese government had established a 20-kilometer radius evacuation zone. Japan did not provide adequate information to the United States, including the fact that it was unable to obtain the necessary information.

While the United States may have been somewhat pushy, Japan should have moved faster in setting up meetings with Japanese officials when NRC officials came to Japan.

Fundamentally, Japanese officials were embarrassed and did not want the U.S. officials to see what had happened. Japanese officials may have also had a sense of pride at being able to handle the situation by themselves.

On March 15, an NRC delegation led by Charles Casto arrived in Japan, and that changed the situation. He made the appropriate judgments and also had consideration for what the other party was going through. That led to an increase of trust among Japanese officials (emphasis added).

On March 17, Japan demonstrated its will as a nation when the SDF [Self-Defense Forces] dropped water from helicopters over two reactors at the Fukushima No. 1 plant. The United States was frustrated that Japan was not employing all the assets that it had, including the SDF. That message was eventually passed on to Kan and Defense Minister Toshimi Kitazawa.

Hosono, former parliamentary defense minister Akihisa Nagashima, and others met with U.S. officials, including Ambassador John Roos, on March 18. The Prime Minister's office made a decision to take the initiative to establish a bilateral body to deal with the nuclear accident on March 22. That led to a more coordinated effort by the Japanese government, although it took 11 days to achieve.

As I have described, over that initial ten-day period, the two governments did not work well together. There was a critical lack of reliable information, skepticism of each other's motives, reluctance to share what little data there was, and worry on the part of the Japanese about a takeover by the Americans. Miscommunications, disorganization, and a mountain of logistical and leadership challenges—how could we work through that? Well, to start with, we set up a series of bilateral meetings.

These meetings revealed the vast—although at times subtle—differences between our respective cultures. In the end, both countries became better organized, more efficient, and much more effective.

When two or more entities must work together to mitigate a crisis, the quicker they can overcome their points of contention—whether they be cultural or corporate—the better. Because once everyone is working toward a common goal, problems are less likely to develop and solidify.

BILATERAL THROUGH THICK AND THIN

Around March 20, Ambassador Roos was contacted by the Japanese government about holding a bilateral meeting with all the players. I think the Japanese were seeking more active help from us and were frustrated by the channels that were in place. To their credit, they recognized that having one primary communications channel—and no backchannels—would facilitate that. On the 21st, we held a "trial" meeting at the Ministry of Defense, with many in attendance. It was not very effective, but it did reassure the Japanese that we were not going to scream, yell, or behave obnoxiously. They requested another meeting to be held the following evening at the Kantei, the principal workplace and residence of the prime minister; in effect, Japan's White House.

The Kantei meeting on March 21 was attended by top-level people from the American government contingent and cabinet ministers from the Japanese government, along with TEPCO and NISA representatives, among others. We met nightly for months from then on.

We'd prepared all day for that initial meeting, and I opened it by expressing my country's deepest regrets about the tragedy befalling Japan, as I always did during those early days. As usual, I also sought one positive note in all the horror, some success or small improvement for which Japan could be congratulated. This wasn't always easy to come up with, but I managed it, and found it helped the tone and mood of meetings and, ultimately, the tenor of our relations.

At that meeting and subsequent ones, after my positive opening after the end of the briefing, the Japanese side would then move to close the meeting without allowing time for questions. Once we'd demonstrated that our intention was not to be critical of their performance, time was allocated for a few questions at the end of the meeting. Gradually, protocol evolved to the point that we could ask questions of each presenter after he had spoken. We had adapted to their meeting etiquette, and eventually they adapted to ours.

I used the Kantei meetings to drive the rhythm of daily team activities. In a rapidly moving crisis, your work rhythm is determined by who you must brief and when. Everything you do during the day feeds into that briefing. As a team leader, I built the team meetings and metrics (in our case, the "Windows Chart") to culminate at the Kantei meeting. Everything I did during the day fed into that higher-level briefing. In any management situation, I recommend the establishment of a work rhythm early on. And in a crisis, the quicker you get that in place, the better off you're going to be.

Indeed, our bilateral meetings turned the tide for both the Japanese and the Americans in terms of getting organized. They became known as the "Hosono Process," named after Diet (Japanese Congress) member Goshi Hosono. Through it, we established an efficient system by which the Japanese could request resources from the many American agencies that had offered help, and we could approve, track, and fulfill those requests.

A SMALL MOMENT OF LEVITY

Although our undertaking could not have been more monumental and serious, several funny stories emerged from the Kantei meetings. Immediately after the first one was adjourned, two Japanese gentlemen wearing earpieces came up to me and asked if I had a minute to talk. "Sure," I said. Clearly, they were the Japanese version of Secret Service members.

"Would you come with us?" they requested, and down we went into the bowels of the Kantei building, through dark, stiflingly hot hallways and into a series of catacombs. We seemed to be winding around and around—I could never have found my way out on my own—and it was obvious we were going someplace top-secret. *Where?*

We emerged in the prime minister's residence.

I met with Goshi Hosono there, and we talked about the status of the reactors—but the meeting was really about nothing more important than showing his eagerness to work with us.

Thinking about it later, it struck me as funny. There I was, a newcomer to Japan, on a secret journey with two sinister-looking strangers. Nobody knew where I was, or even that I was gone. It would be so easy to make me just . . . disappear.

I told myself I'd watched too many episodes of *The Sopranos*. *Just don't let these two guys get behind you,* I thought, only half-jokingly. *Keep them in front of you.* When we arrived at our destination, I could only laugh at myself.

FLAWED SHUTDOWN DECISIONS

Before each Kantei meeting, the whole American delegation would congregate on the eighth floor while the Japanese prepared the conference room on nine. After about twenty minutes, they would call us up and we'd begin.

One evening before we'd even assembled, Hosono came down and directed me and an interpreter into a little conference room on the eighth floor and shut the door. He had his mobile phone on speaker, and it became obvious that the prime minister was on the line. They were discussing the Hamaoka plant, specifically whether to shut it down and how it could be restarted. There was much back-and-forth discussion, which the interpreter translated into my ear.

Hamaoka was a most dangerous site, in that one of the biggest seismic faults in Japan traveled directly underneath it. Kan had decided that it needed to be shut down, which got me thinking about the establishment of criteria for a restart of any shut down reactors. My philosophy at that point was that all that was needed to restart a plant was money. People would want money to make their plants safer, and if there were enough jobs at stake and enough money in a local area, they would be more receptive to the restart. In reality, it is a bit more complicated. In fact, the issue of a shutdown/restart is much like the decision to evacuate we discussed earlier. These are the kinds of decisions that, if made based on emotion and not science, cannot be undone with scientific explanations or technical data. In the end, money alone is not enough and, in fact, is not even the key factor. Public perception of the safety of a reactor restart—and people's acceptance of it—is.

During the years after the accident, Japanese nuclear plants did spend a lot of money to improve safety and security, but this didn't seem to help them get restarted. At that meeting with the prime minister, I knew that if they shut down Hamaoka, the shutdowns would cascade down their entire fleet, because people at each plant would ask, "What about *our* plant? If you shut down Hamaoka, our plant becomes the most vulnerable. Why don't you shut *our* plant down?"

And that is exactly what happened. Once they decided to shut down Hamaoka, public pressure and regulatory concerns led to the shutdown of all operating reactors in Japan.

I am still unconvinced that the issues were considered fully by Kan and others. If they had stayed firm about keeping the plants open, or chosen just a select few sites to shut down (along with the Boiling Water Reactors on the east coast and ocean side), it might have prevented that "cascade" toward a total shutdown. I feel they would have been better off coming up with some very specific criteria right then, rather than deciding based on the vague notion that if there was an active earthquake fault near or under a plant, then it should be shut down. The reality was that *all* the reactors were near earthquake faults—so that vague criterion ultimately led to the complete shutdown of the remaining 44-reactor fleet. Today, in 2018, nine of the shutdown reactors, mostly on the west coast of Japan where tsunamis are not a great threat, have restarted.

The decision to shut down Hamaoka even had repercussions in the United States, which has several nuclear plants—including Diablo Canyon and San Onofre in California—near fault lines. However, the NRC decided that these nuclear plants remained safe, so they remained open.

AVOIDING DOG PILES

At times during the course of working with the Japanese, usually in Kantei meetings—for no reason that we could think of—they would say things that we thought might be nonfactual. I thought of these questionable or "off-kilter" statements as *dog piles* and asked my people to simply "step over them," rather than ask questions or make accusations. Keeping the somewhat silly but apt dog analogy going, my team understood that our success involved avoiding these obstacles in order reach the dog bowl.

To change metaphors, I once used an analogy from the film *The Matrix* to describe our situation. There is a scene in which bullets are shot rapidly at the protagonist but he avoids them by twisting and contorting his body in seemingly impossible ways. "Just let these 'bullets' fly by you," I told my team. There is rarely any point in arguing or debating unless it is on points critical to the mission. The questionable comments that came out in these meetings rarely rose to that level, so I counseled restraint. It was my version of "Don't sweat the small stuff," I suppose.

NEVER LOSE SIGHT OF THE MISSION

Another analogy I used during the response concerned "rabbit holes." A rabbit hole is an endless path that ends up leading nowhere, or back to where you started. To avoid rabbit holes, you must never lose sight of your mission. Therefore, there is real wisdom in asking yourself frequently, "*What is my mission?*" Repeat as needed.

In the case of Fukushima, we had two missions—to protect American citizens and to help the Japanese in any way we could. When issues came up, I asked myself, "How might the resolution of this help protect American citizens?" and, "Does it relate to an area in which the Japanese need help?" Passing the proceedings through the filter of these questions helped me keep irrelevant data from driving my strategy.

I wrote earlier about the dangers of chasing data or allowing yourself to be inundated by it. Almost anything can be correlated to your mission if you don't keep that mission very clear in your mind. "What is my mission?" you must continually ask yourself, "and will the pursuit of this matter contribute to its success?"

NAVIGATING THE BILATERAL MEETINGS

One night, the Japanese started our meeting with a very emotional apology of their own. The prior evening, they had released some radioactive water into the ocean but had failed to notify the international community beforehand. For neighboring countries, this was troubling, to say the least.

The Japanese officials briefed us on the details and stated their regret for not notifying us in advance; then they all performed a *saikerei,* a group apology bow. It was very formal and moving, and conveyed their sincere sense of regret and guilt. I believe that once they became aware that they'd triggered international outrage, they understood their mistake and apologized. We accepted the apology and moved on. I can think of no American cultural gesture that conveys as much.

In contrast, a funny thing happened at one of the nightly meetings in April, when the Japanese Finance Ministry made a very humble request. They struggled to find a polite way of framing it. "Japan is not a Third World nation," they said, "but a wealthy one. . . ." It emerged that they wanted to *reimburse* us for all our help.

It was an awkward request, and we understood that we could not just dismiss it out of hand. We told them that we would make an accounting of the costs and provide them with a "bill." Over the next few weeks, various American agencies and the U.S. military attempted to itemize the expense of the many services we'd rendered to date. The effort almost immediately proved impossible. Now we were the embarrassed party. We couldn't provide a bill.

In the end, we established a date after which costs would be provided to the Japanese as we incurred them. Briefly, it seemed to me as if *we* were the Third World nation.

As an extreme-crisis leader, particularly of an international effort, one must continually exhibit diplomacy, organization, leadership, and humility. The Hosono meetings turned the tide in our communications with the Japanese. If the early ones had gone badly, the lack of cooperation between our governments might have had dire consequences. Poor communications might have led to a more aggressive stance on the part of the United States, and that could have set back the recovery and undermined Japanese public confidence in their government. Poor communications between our military and the Japanese government could have harmed the U.S.-Japan alliance. Although I may not have fully processed it at the time, the pressure on all of us—Americans and Japanese alike—to build a strong bridge of cooperation was immense. The Hosono meetings became that bridge.

CHAPTER 18

KONDO THEORY

---

*Life's failures are only failures if you don't learn from
them and use them to create successes.*

By March 25, things had begun to settle down a little bit and were moving along more smoothly, thanks to the Kantei meetings. That day, Ambassador Roos and an interpreter traveled with me as I responded to an invitation from Goshi Hosono's Lower Diet House office.

Hosono told us that the analysis of Dr. Shunsuke Kondo—considered the father of nuclear power in Japan—presented a catastrophic scenario that suggested the evacuation of many more people than we'd hoped—perhaps from as far away as Tokyo. In fifteen minutes, we were told, that analysis would be presented to the prime minister.

We met with Hosono, Dr. Kondo, and one other person to offer our own version of the worst-case scenario. We would take this opportunity to share with them our worst-case scenario. During this meeting, we shared the results from some of the computer modeling that was created with the "supercore" assumptions. We had calculated that the risk of further damage to the plant was about 1E-3 (one in a thousand chances); we wanted to get this down to 1E-4, (one in ten thousand chances). We believed the way to achieve this was to create a more robust and sustainable independent water system. According to what our risk experts had calculated, we could not get better than 1E-4 risk without adding another injection point. We could add diversity and redundancy to the pumping systems all day long, but without another injection point, we couldn't significantly reduce the risk. We continued to push that point at every opportunity, including this one.

I mentioned earlier that Dr. Kondo had told me his career would have received a failing grade because of Fukushima. Dr. Peter Lyons, the head of the Department of Energy's reactor safety organization at the time, would tell me later that he'd witnessed a huge change in Dr. Kondo's personality after the accident. The respected scientist and academic clearly felt responsible on some level. It was an unfortunate capstone to an illustrious career, which included training most of the nuclear engineers in Japan either directly or indirectly, through his scholarship.

At the meeting, Dr. Kondo shared his scenario and asked us to safeguard it, which we did. When I returned to the embassy, I immediately had the paper classified Top Secret, but, inevitably, it soon leaked to the Japanese media. Hosono asked us about this lapse, but I believe he knew we weren't responsible for the media leak.

The premise of Dr. Kondo's analysis was that a new hydrogen explosion would necessitate another site evacuation, and the resulting abandonment would lead to further meltdowns. The radioactive release would then spread across the Tohoku region to Tokai, and as far as Tokyo. Three nuclear plants, including Fukushima Daini, would then have to be evacuated, and multiple nuclear sites would experience full core damage, spent fuel, and reactor damage. This would trigger the evacuation of a massive area—as far as 108 miles or more—that would include the Tokyo metropolitan area, with more than 20 million people.

I began to call this the "popcorn" scenario.

Dr. Kondo was obviously very troubled by the dreadful implications of this scenario, which I must point out, was predominantly anecdotal and not at all quantitative. It was masterfully laid out, but based solely on conjecture.

For starters, what would cause the hydrogen explosion he described? Based on what I understood about reactor cores, I believed that they had already released all the hydrogen they were going to release. The spent-fuel pools had gone days without water by then, yet they were releasing no radioactivity, and they were continuing to cool the reactors. Dr. Kondo's premise was that if the seawater injection stopped, there would be a further meltdown—but we had evidence to the contrary.

As Dr. Kondo spoke, I imagined how I would respond, what questions I would ask. I was also remembering an incident I had experienced years earlier. On February 14, 2008, the then-NRC chairman, Dr. Nils Diaz, came to our offices in Atlanta. During that visit, I presented him with insights gathered from each of our divisions. After I'd outlined the effectiveness of our reactor oversight process, Dr. Diaz proceeded to tear it apart without mercy. I didn't find his comments technically valid, but I did try to understand why my presentation had failed to impress. As I drove home that night, it came to me—I had tried to teach a professor! What a blunder. Don't ever try to teach a professor anything.

I realized that trying to teach Dr. Kondo would be equally ineffective; in fact, it would be the worst thing I could do. With that in mind, and summoning up my own *listen, learn, help, and lead* mantra, I worked hard to avoid a bossy or pedantic tone as I addressed the expert. I couched my response in a multi-part question.

"Dr. Kondo, help me understand what the operators could be doing that they're not doing now, to prevent your scenario from occurring?" I asked. He didn't have an immediate response, so I pressed on. "I mean, what if they made the injection water system robust enough and automatic enough that the reactors could survive, let's say, ten days without human interaction?" After another long pause, I offered, "Wouldn't that keep your scenario from occurring?"

He reluctantly admitted that if they could automate the systems, this would, in theory, prevent his worst-case scenario from happening.

I then turned to Goshi Hosono and said, "Dr. Kondo just explained the required action." (Not coincidentally, this was consistent with what we had been wanting all along: more robust and sustainable cooling systems.) I suggested that as part of briefing the prime minister, they offer the robust injection system as a counter-measure to Dr. Kondo's scenario.

I believe this marked the dawning of the Japanese understanding that automating the cooling water system and building in redundancy and reliability—particularly of the injection nozzles—would ameliorate the situation. The process of advising and convincing them also took some redundancy, as we revisited these points many times throughout our remaining time there.

Japanese government officials decided that, in the end, if nothing worked and meltdown was imminent, then a slurry mixture of gelatin and sand would be dumped on the spent-fuel pools and reactor head to try to minimize the radioactive release. Based on what we knew about this measure, we conducted some calculations based on the weight of the slurry. We concluded that the mixture would have the opposite of the intended effect—insulating the reactor fuel and causing temperatures to rise even higher. The mix of gelatin and sand did not strike us as viable either, and there were some discussions with a large Japanese nuclear vendor and others about an alternate concoction.

As I told Ambassador Roos, I believed that after the Unit 4 reactor building exploded, the worst was over. But many reporters and others asked about the possibility of a much larger radioactive release if more reactors or spent-fuel pools experienced damage. Had more cores been compromised, there would have been a denser plume with higher isotopic concentration. Where and how far it traveled would depend on the wind, which could not be predicted with any accuracy. But, contrary to popular notion, three reactors melting down would not create three times the radioactive plume, because the calculation of such a thing is not linear; it's logarithmic.

In the years since, when people ask me what the worst case might have been, I say that it wouldn't involve more distance but a higher concentration of radiation. There is an important point to make here: Even the experts disagreed about the radiation impacts, and ultimately, the actions taken to protect the public were based on miscalculations. In the end, the science of how widespread an evacuation should be and what the short- and long-term health effects are on those in the path of the plume are debatable. Unfortunately, in any case in which evacuations are ordered, innocent civilians are likely to lose homes, jobs, and their sense of safety. In that sense, unnecessary evacuations are almost as harmful as necessary ones.[1]

Radiation experts often disagree about the impact of exposure to humans. In recent years, the International Atomic Energy Agency (IAEA) has modified its standards for reentry into a contaminated zone from a strict science-based approach to a more local community decision. That is a positive change. I believe the best course of action is to provide local communities with the data and let the locals decide for themselves when and if to reenter the zone.

This is an important point: No one—not even the plants' operators—died from overexposure to radiation at Fukushima. (Some critical-care patients died during the evacuation, but reported numbers vary widely.)

As an extreme-crisis leader, you may face situations in which credible and estimable individuals raise troublesome issues. It is up to you to dispense with those issues. In this case, an experienced and esteemed scientist took a scenario to the prime minister, and it had to be addressed. Drawing upon my past experiences, I addressed the issue in a palatable way and simultaneously endeavored to solve it. I sum it up this way: Life's failures are only failures if you don't learn from them and use them to create successes.

CHAPTER 19

J-VILLAGE AND FIRST TRIP TO DAIICHI AND DAINI

There were mountains of contaminated materials over fifteen feet high, covering the range of an international soccer field: facemask filters, Tyvek protective suits, respirator filters, you name it. This was the radioactive waste from the thousands of people working at the site and traveling to and from it.

Extreme-crisis leaders know that they must visit the crisis site to survey the situation for themselves, understand it, feel it in their marrow. I did my doctoral dissertation on this subject, and my research supported the fact that the farther away a leader is from the event, the less involvement and emotional attachment leaders hold. In an extreme crisis, the brain of the leaders must be fully engaged, yes, but their emotions are key as well.

Devising an extreme-crisis response, it is important to understand that the street-level leaders may be fearful for their lives. It's vital that the top leadership understand the consequences of the event: how it's affecting the citizens, responders, and environment. Even in emergencies that are not life threatening, people still have fears that must be understood: for their livelihoods, ongoing working conditions, and so on.

The first time I heard of J-Village was in the organizational chart that the Japanese presented to us in the first week of the accident. The chart showed the Japanese Ministry of Defense in charge and on the scene, and it showed the placement of the ERC and a staging area called J-Village. A soccer training facility located 14 miles south of Fukushima Daiichi NPP, J-Village was transformed into a staging area for workers entering and leaving the nuclear power station. They put on their protective equipment there and performed decontamination examinations to ensure that radioactivity did not spread.

Admiral Tom Rowden, our Navy attaché and Nimitz Strike Group Commander,[1] and I were in the process of making plans to go to Daiichi. Early on in our Kantei meetings, Tom had pushed to get clearance for four of us to visit J-Village, but there wasn't a lot of love for that proposal. Perhaps the Japanese leaders were concerned that the people there were too busy to entertain international guests, or, more likely, that the visit would set a precedent of allowing foreigners in. The simplest explanation was that they did not want us to witness the drama unfolding at J-Village.

At some point, through back channels at the MOD, Rowden went alone by helicopter to J-Village. He came back that night and described to those of us at the Kantei meeting what he'd seen. He explained that it was a massive organizational rallying center, and expressed that he was very impressed with what he saw there. Everything we needed to know about the emergency response could be seen at J-Village, he reported.

While he believed that TEPCO was commanding all on-site activities from J-Village, I had my doubts. It was my firm belief that the operators themselves were in charge of managing the crisis, and I would have been shocked to learn otherwise. My inclination was not to lean too far forward in congratulating the leadership at J-Village. "There's still a long way to go with this response," I cautioned. "It's good that they're making progress up there, but we still have a long way to go."

Admiral Rowden urged the Japanese to permit me to go to J-Village myself, and ultimately, they approved a short trip—but we were told not to visit the site itself. The next day, armed with personal radiation measurement instruments, Tom and I flew north to J-Village.

From the helicopter, I could see the widespread tsunami and earthquake damage and significant infrastructure destruction. The many homes with blue tarps covering their roofs made the countryside look like a checkerboard. I didn't initially grasp the reason

for these, but I found out later that during earthquakes, loose ter-
racotta roof tiles slide down and take other tiles with them. This
was the cause of damage to tens of thousands of homes in the area.

I also noticed that buildings under construction or renovation,
even smokestacks, were wrapped with fabric. This struck me as
more of an "optical solution" than anything, and not particularly
effective for safety purposes. *If I can't see it, it's not there,* was what I
termed this sort of measure. This was a phrase I thought quite apt
and pithy, but it took on a darker meaning when a useless temporary building was constructed over Daiichi Unit 1. The ineffective
helicopter drops of water onto the spent-fuel pools constituted a
similar kind of effort.

Seeing the tsunami and earthquake damage from the air was
startling, dumbfounding—almost incomprehensible. Looking
down at the remnants of entire villages that had been swept away,
along with their inhabitants, made the events real to me in a new
way, and the power of this sight only intensified on later driving
trips through the area.

The Japanese took a lot of pride in their commandeered soccer
facility, but we could see from above that the hundreds of military
vehicles, trucks, and cars were in the process of ravaging the
grounds there, making deep ruts and wreaking havoc. The pilot
circled to check out the landing area, and when we were about 50
feet off the ground, the alarms on our radiation devices went off.
This was alarming, especially since J-Village was a full 14 miles away
from the nuclear plant itself. I'd never encountered a situation
where radiation levels were so high at that distance. Without exchanging words, we silenced our devices and pressed on.

I thought back to that anomalous radiation reading that had
been reported. Maybe it had been correct after all! Partly because
we had a radiation technician with us, we felt comfortable in raising
the alarm-trigger point on our own measurement devices to above
the "background level" we'd noted. This way, we'd be alerted to any
further increases in radiation.

As we got out of the helicopter and walked toward the J-Village,
I was struck by the number of satellite trucks and military vehicles,
as well as the enormous amount of equipment and detritus. There

Figure 8. Temporary evacuation site housing in Fukushima Prefecture

were mountains of contaminated materials over fifteen feet high, covering the range of an international soccer field: facemask filters, Tyvek protective suits, respirator filters, you name it. This was the radioactive waste from the thousands of people working at the site and traveling to and from it.

It was March and still cool outside in northern Japan, and there was no water or electrical power in J-Village. The site was as unsanitary as can be imagined. No women were allowed in for this reason, and because of the complete lack of privacy. Lines formed in front of each of the ten or twenty reeking portable toilets. Men changed clothes out in the open. The only water to be found was bottled water, which was a bit precious to use to clean oneself.

Just outside the facility, an empty bus sat on the side of the road. As we passed it, my radiation device went off. The bus was empty, but it had been contaminated heavily as it conveyed workers back-and-forth to the site. Contamination is so insidious that way—it travels and spreads via people, the wind, vehicles, you name it.

As we entered the building—which was just as cold inside as out—we were greeted by TEPCO employees and members of the SDF. The odor inside overwhelmed us. All around us was seeming chaos: people trying to sleep, eating, talking, recovering, in various stages of undress, lining up to travel the eighteen miles to the damaged reactors. Some areas had been tented off for the leadership teams, or for viewing video feeds from the helicopters. When I entered one of these, several people were watching a helicopter circling Daiichi taking pictures of the damage including the spent-fuel pools.

We had been under the misapprehension that Daini was safe because it had a working emergency diesel generator for electricity, but thousands of people were dressing in protective clothing to go there. We learned that they'd started requiring respirator face masks for the workers traveling to Daini. I viewed this as an over-abundance of caution. I had been aware of the inundation at Daini by tsunami water, but hadn't realized the real extent of the damage until that moment. It was so much worse than I'd anticipated—a major disaster in its own right. Ambassador Roos's wife Susie had been absolutely right to ask, "What about Daini?"

Ministry of Defense and TEPCO employees briefed us about the conditions of Daiichi. From what I could see, though, J-Village was merely a staging area; no one there was conducting any strategic leadership activities, as we'd been led to believe at the Kantei meeting. Other than logistical leadership, I could see no one working on ways to resolve, mitigate, or address the accident itself. My preliminary conclusion that we were overestimating the leadership activities at J-Village proved to be accurate.

A thorough tour of J-Village revealed two notable things. First there was the lack of oversight and regulation of health and contamination practices. A significant amount of cross-contamination was occurring between contaminated "hot zones" and uncontaminated "clean" zones. Second, no leadership or command center appeared to be directing activities.

The cross-contamination was occurring throughout the entirety of J-Village. Thousands of workers were traveling to and from the sites, but as far as I could discern, the extent of their precautions against contamination amounted to using one side of the highway as "hot," and the other side as "clean." Needless to say, I had lots of questions about all this, but couldn't get any answers.

We'd faced this issue after accidents in the U.S., and I knew that managing it is an enormous challenge. In the wake of Fukushima, I understood that a "macro approach" seemed to be the best they could do, given the vast numbers responding. As the months passed, the controls did grow more stringent, and steps were taken to limit and monitor the spread of radiation. Eventually, they began decontaminating everyone—but this took months to put into effect. We left J-Village with quite a lot to report at the next Kantei meeting.

MY FIRST TRIP TO DAIICHI

Several weeks later, some of the NRC team, including me, traveled back to J-Village—where conditions had not changed—and then to Daiichi. Quite simply, traveling through the evacuated zone on a foggy and rainy day took my breath away. The windows inside the van kept fogging up and we would wipe the condensation off so we could see the massive damage to the countryside. It wrenched our hearts to see the destruction, both natural and manmade. We moved from side to side of the vehicle like children: "Look at this! . . . See over there!"

It was extraordinary to see significant destruction next to intact buildings unaffected by the tsunami or earthquake. Tragically, a blanket of radiation had hit some of the areas spared by nature. Very few people were unaffected by some aspect of this "triple disaster."

We saw where the tsunami had blown over ocean walls meant to prevent landfall, the water flowing into villages, through valleys, and around mountains. We saw where it had swept entire villages away, leaving only one-foot-high foundations and scattered possessions. Mounds of such detritus had been piled up by bulldozers so that emergency vehicles could get through.

Our TEPCO escort was using a radiation detection meter, and every few minutes, he would call out the level of radiation. His booming voice made the trip very unsettling. As we went through the villages, the radiation level would go down; in forested areas, it would go up. This told us that rain had washed away much of the contamination from the hard surfaces of the populated areas, while it remained in the natural landscape.

The only vehicles we saw were TEPCO or military. The abandoned roads were heavily damaged, and we had to take several detours because of sinkholes filled with cars! *What happened to the people who had been in those vehicles,* I couldn't help wondering. I saw a Toyota sticking up out of a crevasse and dreaded the thought of what its occupants must have experienced as the road surged up in front of them.

I saw a parking lot full of Lexus cars knowing that they were coated in cesium by the radioactive plume, and understood that they would have to be destroyed. The magnitude of the assets lost by this country seemed inestimable at that moment. I thought about the unprotected factories. Had there been enough warning, they could've shut down their ventilation systems, and perhaps by sheltering, preserved their assets. But that was not the case.

The human impact was staggering. The front doors of some of the abandoned homes remained open, all manner of personal possessions scattered about the yards. As we traveled slowly through a small village, I peered into a coffee shop where coffee cups still sat on the counters. People had simply fled with nothing.

The government was endeavoring to keep people out of the area, for safety reasons and to prevent looting. It was as if the area had been attacked by an invisible enemy—and in fact, it had. There had been no evacuation zones at Three Mile Island or Paks, while the zone at Chernobyl had been the same size as the one we were driving through.

Every level of government was stressed to the breaking point, the task of restoring an entire country mindboggling. Among the thousands of lives that had been lost to the tsunami were innumerable first responders, health professionals, and government employees, and at the onset, it was the dearth of first responders that had caused the most difficulty. The plan for tsunami disasters is to dock hospital ships at the ports as temporary medical facilities, but in this case, the wave had knocked out the ports.

As we moved deeper inland, I started to be able to distinguish between tsunami and earthquake damage. Collapsed facades were likely due to earthquake damage; where facades were gone altogether, along with any debris, it was probably the work of the wave.

And, much like the devastation of a tornado, tsunami damage is concentrated in certain areas.

At about the 6-mile mark from Daiichi, our van stopped. Up to that point we had been wearing paper facemasks to protect us from airborne radiation, but we now switched to our full-face respirators. There wasn't much need to move off the road to stop because there were no vehicles around, but we pulled into an abandoned gas station anyway. My thoughts turned to the people who would have been here on the morning of March 11, pumping gas, going into the small convenience store, doing the same everyday things we all do without thinking. They had been much like me on that same Saturday morning, getting gas at a local Walmart. Now, there wasn't a soul around, other than a few dogs, no doubt looking for the humans who would normally feed them. It may have been weeks since they had seen anyone. The shelter facilities would not accept animals, so residents were forced to leave them to fend for themselves. The devastating aftereffects felt endless.

In a surreal touch, we heard about an ostrich loose in the area. Supposedly, the locals had named it "TEPCO" to chide the utility for sticking its head in the sand.

BLEAK FUTURE

As we drove, we spoke about the complete loss of the current rice crop and the bleak future for local farmers, who would also have to destroy their livestock. Out of desperation, or unknowingly, a few farmers had snuck their horses, cows, and pigs out of the area undetected. Those animals would be responsible for contaminating yet more territory. Very near the plant, I saw a bridge whose ramps had fallen away. There was no way to get onto it. An old roadway that ran parallel to the road we were on had collapsed as well.

At last, we arrived at the Daiichi site, and the amount of damage was unfathomable. The photos and videos we had seen—most taken by aircraft or satellites—did not begin to communicate the devastation. We drove through the site speechless, just soaking it in. Major buildings destroyed, steel twisted by the force of water, pumps lifted off their bases and moved hundreds of feet, massive cranes toppled. *How did water twist those beams?*

On the ocean side, the destruction was total. The forces had flung vehicles everywhere; they were jammed inside of buildings, turned upside down, crushed. Though it didn't seem possible, the explosion of the reactor buildings had caused even more damage. When Unit 1 exploded, the entire side of a maintenance building *a full mile away* was shorn off and the remaining windows were pulverized.

The office buildings on the site, including the engineering building, were decimated by the force of the blasts. There was no electricity anywhere and contamination everywhere. Records and technical information critical to the restoration of the plant were simply gone.

Several things struck me as I surveyed these calamitous scenes. First, we must do everything in our power to prevent a release of radiation from happening in the United States. Second, I wondered if even the deployment of a nuclear weapon would cause as much damage as this combination of natural and nuclear disaster. *How many nuclear weapons would it take to cause this kind of damage?* I made a mental note to try to calculate the force of the tsunami, to determine the answer to this awful conundrum.[2]

Our first stop at Daiichi was the ERC, where Yoshida and Inagaki were leading the operators through the accident. We found the entryway and were greeted by pure chaos. Dozens of people were standing outside an airlock door, trying to maintain the airlock and prevent contamination from entering the building—essentially a useless exercise. Stacks of contaminated dress-out equipment, drums, and waste were everywhere. Trying to isolate the contaminants was nearly impossible under the conditions there.

Inside, the odor of unwashed bodies, unclean clothing, and rotting food was pervasive. There was no running water in the facility, no working restrooms. People occupied every square inch available— swarming in stairwells and closets, on floors and atop desks, sleeping on plastic sheeting, recuperating, or waiting on an assignment.

Officials from TEPCO and Japan's Self-Defense Forces briefed us in a small conference room, then we made our way up the stairs to the main control center, climbing over people as we went. We found a very sizable command center with a video screen on one wall and a work area shaped like a horseshoe. Here, too, workers were everywhere; in the back, they had tried to organize sections for resting, eating, and yoga.

There didn't seem to be any place to find genuine solace other than, perhaps, one wall, where good wishes from around the world had been posted. Schoolchildren had written letters and drawn pictures to encourage the operators. Someone had assembled a shrine. There were bundles of origami cranes hanging everywhere, including the 5,000 cranes from the NRC that I had delivered to Hosono a week earlier.

We went to the front of the room where I met Yoshida and he briefed me on the status of the reactors. He seemed like a gentle giant—a very tall and lean man, very friendly. One of the astonishing things he told me was that they had survived on just one small portion of rice and one bottle of water per day for at least the first week (see Figure 9). Each person had to decide how much of his water to drink and how much to use to boil his rice. At one point, the workers were asked for their cash so that food and water could be purchased for them.

Yoshida showed us a single, remote feed on a laptop computer trained on a gauge in the reactors—they were using this as a (very rough) measure of reactor water levels. That was pretty much the extent of the instrumentation they had.

In the ERC, most of the people working at the tables to try to restore systems and components were either systems or component engineers, not plant operators. In the United States, when we have emergency drills, the operators are the key responders. In this particular accident, however, where they were without instrumentation or energy to power the controls, it didn't make sense for the operators to be involved on the front line, at least not yet. When I asked to talk with some of the operators, it was a struggle for them to find any. "Oh yes, operators. We have some of them around here. I think that they're back there," someone told me, gesturing *that-a-way*. I found them seated in the back of the control room, waiting for an assignment, and met with several of them.

Figure 9. Daiichi Plant Superintendent Masuo Yoshida with rice and water

The operators spoke about their losses; I think seven of them had lost family in the tsunami. I asked them in turn how long they'd been working there, and many had been there for twenty years. One very nice gentleman had a bit of a potbelly that mirrored mine. Someone snapped a photo as we stood comparing them. "This isn't a potbelly, it's a fuel tank for a love machine," the man quipped, and it brought the house down.

Figure 10. The author (center left) with Yoshida (center) at Daiichi (source: TEPCO)

Yoshida rolled with laughter. It was a rare bit of levity we all needed.

We went on to tour the plant, and I can barely put into words my feelings as we approached it. Right away, we saw the damage to the administration building: Its windows were gone and it was fully contaminated. I was told that they had tried to work in that building for a while, but the roof had been on the brink of collapse so they abandoned it.

I saw 100 or more firetrucks that were being used a few at a time to feed water to the reactors. Many of them were very old, which struck me as odd until I learned that local fire departments had chosen to sacrifice only their oldest trucks. What's more, they would not even drive them to Daiichi, for fear of radiation exposure. This meant that the plant operators, who were unfamiliar with the operation of firetrucks, had to drive out to pick them up. When a pump on one of them gave out, they simply abandoned it.

As we drove around the periphery of the plant, the damaged reactor buildings came into view. Endless fire hoses, probably what amounted to hundreds of miles of them, were snaking through the grass, running up the sides of tanks, and traversing the roadways.

It hit me hard: This was the heart of the pumping system keeping the cores covered, and there was no method or control being exerted at all. This was a seat-of-the-pants operation all the way. All of the engineering, planning, procedures, and operations meant to maintain viable reactor cooling systems had come down to this: fire hoses. Unbelievable!

I noted immediately how vulnerable the whole water-delivery system was and that there was no management or oversight of the hoses whatsoever. There was nothing protecting them from being run over, destroyed by another blast, or washed away by another tsunami. I worried about them rupturing, which might halt the flow of water and cause highly radioactive water to spill out onto the ground. This in turn might limit access to the plant. On top of everything else, these were single-wall hoses without water-tight connections. The fact that there was no protection, no resilience, no redundancy in this system affirmed our determination to bring in as robust a pumping system as humanly possible to replace it.

Continuing down the road, we saw radioactive debris spread across many areas. Rather than remove it, they had sprayed fixative on buildings, trees, and the ground to keep the contamination from dispersing farther. At one point, roughly half a mile from the reactors, the van stopped and we got out to take some pictures. I got very close to a piece of contaminated debris—too close—and it shook up our escorts. Exposing me to radiation on their watch wouldn't be a good thing.

Nevertheless, we all gathered to take in our first view of all the reactor buildings. There were massive concrete blocks hanging by rebar and debris on top of the structures. Frankly, I was incredulous at the sight, which was made even more eerie by the drizzly weather. It appeared as if the reactors were rising out of some ungodly mist. This dreadful view was topped off by the steam and smoke still rising from the fuel pool of Unit 3.

One perspective offered a now calm ocean, its gentle waves lapping at the shore. But when we turned our heads slightly, the devastated plant came into view. The contrast was dramatic. Was it too fanciful to imagine a lesson in this? The natural world had returned to normal, but the machines of man's devising would probably never come back.

I would not completely process all that I had seen that day for some time, and honestly, I didn't know how to put my impressions into words. What we were seeing represented the most devastating challenge humans had ever faced by the dual forces of nature and physics. America had endured the great San Francisco fire, Hurricane Katrina, and the Three Mile Island accident—but not all at once. The people tasked with resolving those catastrophes had struggled mightily to do so, and the situation at Fukushima was three times the intensity of any one of them. It was not apparent to me how the Japanese would be able to find the wisdom and super-human courage necessary to prevail.

We left our perch and drove around a warehouse building near an electrical power switchyard, where a transformer seemed undam-aged. The warehouse building nearby, however, which was a mere half mile from Unit 1, had been leveled. We were told by our escorts that the Unit 1 explosion had caused damage as far as a mile away. The transformers had been saved by their concrete buttresses, but debris covered every inch of the landscape. Clearing it was on the to-do list, but there were so many other things to do first.

We went down a hill and caught a view of Unit 4. There are no words to describe what we saw. The back of the building was in ruins and the front—on the ocean side—unimaginably worse. Vehicles had been jettisoned into the building in every kind of position. The auxiliary boiler was wrecked and steel beams were twisted like pipe cleaners. Again, I wondered, *How could water wreak such havoc?*

We drove on to the seawater intake area over what was now ex-tremely pockmarked terrain. As it came into view, it was apparent that nearly all of the pumps and other components that normally supplied water to the reactors were simply gone. I tried without success to visualize an event that could uproot twenty or thirty mas-sive machines and wash them away, along with the surrounding buildings. Only small pipes and equipment that didn't offer a lot of surface area to the ocean remained.

THERE ARE NO WORDS

We circled back to Unit 4 and drove even closer to the other reactor buildings. The reactors had essentially been demolished (see Figure 11). Again, the perspective that I got from seeing the devastation up close and in three dimensions was much more disturbing than anything I'd seen via satellite imagery.

In spite of my deep sense of shock, as a professional in this field I tried to observe everything with an analytical eye. The structural integrity of the reactor buildings, with their elevated spent-fuel pools, was a significant concern. Because they were forty feet above ground, I wondered if they could withstand another major quake or tsunami. This did not seem like idle speculation as we were still experiencing frequent aftershocks.

Figure 11. Daiichi damage. Source: Tokyo Electric Power Company (http://photo .tepco.co.jp/library/110316/110316_1f_chijou_1.jpg).

We moved on to Units 5 and 6 and looked at a shared spent fuel storage building that the tsunami had filled with seawater. Seaweed festooned the interior. Our escorts told us that the loaded reactor fuel casks, weighing tons, hadn't moved during the tsunami, but that some of the empty casks did. Units 5 and 6 had not sustained any nuclear-related damage, but the earthquake and tsunami had caused significant damage to their outer buildings.

An SUV had been tossed into Unit 6, where it came to rest on its nose behind some transformers. Many, many private vehicles had washed up against hillsides; some stood with open hoods, having had their batteries harvested by operators in an effort to repower the reactor instrumentation.

We walked into Unit 5 to see where the tsunami had washed through the building. Seawater remained trapped inside the turbine, as there was no power to pump it out. It was eerie navigating through the dark and deserted building, its contents flung into every position imaginable. Anything and everything that had been loose inside the building had been violently rearranged and draped in seaweed.

What I saw next absolutely floored me. The flood had swept away the radiation portal monitors and detection equipment. This would cause a colossal problem in measuring exposure of the workers. I was making constant mental notes of issues to tackle and this was certainly one of them.

We descended to the basement of Unit 5, where we could see the trapped seawater. I knew the situation was similar in Units 1 and 4, but the radiation levels in those buildings were too elevated for us to enter. The radioactive water storage problems were evident—and my "to-do" list grew longer. Next, we went down to the ocean side of Unit 1, where barges had been used to move equipment to the plant. All the barges were gone, carried off by the tsunami.

We ended up back on the perimeter road, passing a toppled off-site electrical power supply tower that looked like a jagged piece of origami. The earth underneath the tower had liquefied during the earthquake, causing it to collapse. That single electrical tower had been crucial to the survival of the plant, and a workaround would be needed.

From a safety perspective, the biggest task for the Daiichi leadership became the treatment of the highly radioactive water. Early in our response, we'd had debates regarding the amount of groundwater flow from the mountains that was coursing through the plant headed for the ocean. The problem was that this fresh water would become contaminated by the fuel on its journey to sea. The Japanese did not believe that this was of concern, but our hydrogeologist calculated that it was significant. Ultimately, everyone recognized that the groundwater flow was of consequence and we needed to address it.

Our analysis suggested that an underground wall should be built on the mountain side of the plant. We also suggested the drilling of "extraction" wells to pull the water off the site. Those ideas were deemed too expensive. The Japanese wanted to build the wall *in front* of the plant, between it and the ocean. Not only was this less expensive, it would be visible to the local fishermen who were demanding action regarding the radioactive water. It became another "optical solution."

We had many discussions about building a land-based, ocean-retaining wall in the hope that it would limit ocean contamination. Mr. Sumio Mabuchi, a Diet member and former minister who oversaw transport, also administered the retaining wall. People would say of Mabuchi that he was so tough he walked down the street with a bridge under one arm and a dam under the other!

TEPCO was very worried about sealing off Unit 4 from the ocean because its turbine building was the closest to the shore. They figured it would take around 400 days for the radioactive ground water to reach the ocean. We talked with Minister Mabuchi several times about building a seawall on the mountain side. Our technical people had concluded that if the wall were built on the ocean side first, it would collect the groundwater from the mountains and flood the site even more. We also raised the need for extraction wells around the buildings and "sentinel" (early-warning) wells to detect any radioactive contamination.

Mabuchi and his team ultimately did take our advice and put in some extraction wells, but it took far too long to come to agreement about this and other urgent issues. Over the next few years, the Japanese installed a seawall, drilled extraction wells, and constructed an "ice wall"—an actual wall of ice around the plant. Collectively, these solutions would reduce the amount of groundwater by two-thirds, though the matter still remains a challenge. As I looked around the site, I was alarmed by the multitude of sources of radioactive waste streams. The Japanese were storing some waste in single-shelled tanks, as we had at Hanford[3] and elsewhere in the 1950s, as part of our efforts to build nuclear weapons. (We are still cleaning up those sites!) *What is the long-term prognosis for a site with so much radioactive water?* I wondered. As always, I encouraged the Japanese to think of long-term strategies. Perhaps they should build another facility nearby that could merge and consolidate all those waste streams. It seemed to me they were creating another Hanford, and the problem remains as of this writing.

To limit radiation leakage and prevent ocean contamination, the Japanese installed "silk curtains" in the one mile ocean intake. Incredibly, they did a great job at accumulating the effluent—the isotopes—but the isotopes then dropped onto the ocean bed near the intake. We all knew that, at some point, this too would require a major cleanup. As is frequently the case, one solution causes new problems. All leaders face difficult decisions requiring the balance of both short- and long-term solutions and the problems those solutions cause.

From the highest point of our tour, we could see hundreds of storage tanks under construction that would eventually be used to store highly radioactive water. As I observed the process, I saw that some of the tanks had large berms around them to protect people from the radiation that would be emitted by the tanks. Rubber connections joined all 600 structures, essentially making them into one giant tank. I was concerned about everything, including the degradation of the rubber connections, which would cause leaks of highly radioactive water. (Later, these tanks would be replaced with welded-seam versions, which were much less susceptible to leaks.)

The challenges kept mounting. At the control center for the water treatment facilities, the storage tanks were giving off a significant amount of radiation. Even though its control center was a bunkered facility, the people operating it were quickly approaching their radiation dose limits. Measures were taken to filter and evaporate the amount of highly contaminated water on site. These were quite successful in removing the material from the water, but the contaminated filters became a solid-waste problem in their own right.

In any case, one contamination problem could not be solved. Tritium—a radioactive atom of hydrogen—is very pretty much impossible to remove from water. The good news is that tritium is not a real health hazard to humans. Thousands of tons of tritiated water are still stored at the site. While many governments have agreed that TEPCO should release the tritiated water, the fisherman of Fukushima Prefecture are dead-set against it. At this point, seven years after the accident, the controversy still rages.

As we drove away from the site and traveled back through the evacuated area, we remained speechless. Although the tour had drained everyone of energy and it was getting late, we took the seashore route so we could see more tsunami damage along the way. Each of us just sat back and soaked it in—the innumerable towns and villages emptied of life, the homes and businesses abandoned, infrastructure destroyed, lives gone.

How did anyone survive the horrors that passed through this place? For eighteen miles, we saw very few humans; the area was desolate, nightmarish. Where there had once been thriving communities, nothing remained. It was very emotional for all of us.

I thought about the first responders. As I've said, these heroic individuals were at a premium. Many had been lost during the disaster, and others had left to focus on their own displaced loved ones. Because first responders are key to a successful recovery, Japan's Self-Defense Forces and ultimately the U.S. military were called on to augment their ranks.

VISITING DAINI

I began looking for an opportunity to swing by Daini, and a visit by Dr. John Holdren, President Obama's science advisor to both Fukushima plants, would give me my chance. I would escort him to both Daiichi and Daini.[4] As we approached Daini from the road, we had not seen any discernible damage, nor did anything seem amiss from the mountainside. On arrival, we went to a conference room for a briefing from the staff. They told us the tsunami had struck them hard and inundated much of the plant, particularly the intake structure. After the overview, Plant Superintendent Naohiro Masuda and his staff took us through the nearly deserted site to show us the damage firsthand. The silence was palpable as we navigated the empty hallways to the diesel building. In several places, we saw portable pumps working constantly to expel standing water. Seeing the water damage, sand, and seaweed everywhere—even up on the rusting light fixtures near the ceiling—was eye opening. For the first time, it hit me, viscerally, that something unspeakably destructive had passed through here and how incredible, how fortunate it was that no one had died on site and that the reactor cores had not been damaged.

As we approached the buildings housing the emergency diesel generators, we could see the devastation that hadn't been visible from the road. Ironically, the road down to the ocean had served as the "path of least resistance" for the tsunami, funneling the massive wall of water up toward the reactor buildings and the vital emergency diesel generator rooms.

The tsunami water had come up from the seaside and traveled across the road and around the back of Daini; it had then poured down through the intake louvers of the emergency diesel generators, through the air-intake fans and into the below-ground-level diesel rooms, filling them with seawater. The damage was extensive, but employees were already hard at work rebuilding the emergency diesel generators and cleaning the rooms.

The relative isolation of the Daini site hampered the recovery effort. Its top priority—as at Daiichi—was to restore full electrical power to cool the reactors. Hundreds of workers labored in concert, employing a helicopter to lift reels of electrical cable, then connecting the cables from a radioactive-waste material processing building to the reactors about one mile away. It became clear to me in studying what transpired there that if it hadn't been for the dramatic events of Daiichi, the world would be talking about Daini and the heroism of its dedicated employees.

When I questioned Masuda about the procedures they'd followed, he told me that there were some general guidelines, including the Americans' Severe Accident Management Guidelines (SAMGs), but that these provided few useful specifics. It had been up to him and his staff to navigate the disaster as best they could, based on their instincts.

In the immediate aftermath of the disaster, the staff had little or, at best, problematic contact with the outside world. They did manage to arrange for several truckloads of replacement equipment, but a lot of it was lost in the mountains enroute because the truck drivers didn't know where they were going and couldn't receive directions.

Masuda described how they had stopped work and sought shelter each time a radioactive plume from Daiichi approached the site. Exposed as they were to the fallout from their sister plant, they kept on working, just as the brave American soldiers and sailors of the Seventh Fleet had done when the USS *Ronald Reagan* and its support ships sailed through the plumes.

After my second trip to the plant—during which I was struck yet again by the hard work and resourcefulness of its staff—I attended a status meeting with American Ambassador Roos. When it was my turn to speak, I paid tribute to the people of Daini. As I spoke about what I'd seen there, one of the TEPCO executives in attendance broke down, clearly moved by my praise and appreciation of his team.

In thinking about the courage exhibited in those dark days, I came up with a new goal for the nuclear power industry—never to need heroes. Our engineering design, procedures, and equipment must be robust enough, safe enough, and protected well enough that we simply won't need them. We remain proud of all our heroes—including those of March 11—and we honor them. *But,* I think to myself, *in the end, it would be infinitely better if there were never a need to honor this kind of sacrifice.* It would mean that our industry was inherently safe. There is no higher goal than that.

CHAPTER 20

LITTLE TEXAS

I was committed to my goal from Day One, but my
experiences with the people of the affected area deepened
my commitment. They became a part of my life, and I
began to love them just as I do my native countrymen.
That affinity drove everything I did from that point on.

As an extreme-crisis leader, it is important to understand how the crisis is affecting real, everyday people. Toward that end, I was fortunate to spend some time in a local bar in Tokyo called Little Texas and speak to real, everyday citizens for the first time. I came to understand what they were feeling, empathize with them over their worries, and truly listen to them across the spectrum. What I realized from the unique experience was that cultures may differ but people are people—and looking them in the eye as you listen with your heart can create genuine kinship. Particularly when you are working in a foreign culture, keep in mind that there is nothing like interacting with regular citizens—really hearing them share their thoughts—for gaining perspective on the specific challenges you face.

There are plenty of things to do in Tokyo, but I was fortunate to have trusted advisors in Suzanne Basella and Matt Fuller, who enthused about the cozy country bar called Little Texas. We'd talked about going several times, but were too busy in those first few weeks to do so.

By the time we felt we could get away for some much-needed R&R, I was exhausted. But Matt, Suzanne, and John Basella coaxed me, and it seemed like the perfect opportunity to spend time with them.

A Texas flag at street level clued us in to the small basement dive below. Inside, the place was decorated wall to wall in Austin, Texas, paraphernalia. It was a rowdy room, packed with Japanese people who, inexplicably, seemed to love Texas. A Japanese band was on stage performing American country music. (When attempting to talk with its members between sets, I discovered they didn't speak a word of English, but were singing phonetically by rote—and they sounded amazing! They informed us that they always close with "America." They also confessed that they wanted to visit the States—particularly Texas.)

By the time I caught up with my friends, they were already in good spirits. The cost of a bucket of five decent light beers was fifty dollars, and the bar food was pricey, too, but it felt great to relax. A line-dancing teacher named Naomi had people up on the floor, most of them decked out in incredible "cowboy bling."

When Matt and Suzanne informed people in our vicinity that I lived in Texas, I became the most popular person in the room. Dozens of people came over to the Texas "expert" to thank me, and I wasn't quite sure whether they were thanking me for being Texan, for being a nuclear expert, or just for being American. Women came over and kissed me on the cheek or bowed politely, as if I were personally responsible for the good things America had done for Japan.

It was my first real-life interaction with non-TEPCO or government people, and it definitely broadened my perspective. The disaster had affected everyone in such a powerful way, and their behavior toward me underscored the genuine appreciation the Japanese seemed to have for anyone willing to come to their aid. It was clear that they didn't fully trust their own government to handle the crisis and had pinned their hopes on the "professional" help we offered.

Our second visit to Little Texas, a few months later, was decidedly different but just as memorable. At a certain point, I became vaguely aware that the entire bar had begun playing a game of rock-paper-scissors. I began randomly throwing one of the three signs, not particularly worried about whether I was winning but wanting to participate. Suddenly, someone tapped me on the shoulder and told me that I was in the top ten.

At that point, I began paying a bit more attention, but I still wasn't very invested in the game. After another round or two, the same fellow grabbed me and said, "You're in the top five. You need to go up on stage and play!" I followed orders and when I got to the stage, four Japanese men were waiting for me—my competition, apparently.

We'd all had a few drinks and Matt and Suzanne were whooping and hollering for me. They were regulars, and Matt was a legend for his dancing. "Chuck, if you don't win this thing," he had said to me as I made my way up front, "you can't come back to the embassy. You know that, right?"

Perhaps it was the beer, but I was feeling pretty confident. The grand prize was a set of longhorns and I was determined to win it. I decided that I would play *paper*. It sounds a bit crazy now, but I had a logical reason. I figured that the other men—being men—would want to show off their masculinity, and would all choose *rock*. If I was right, my *paper* would win the day—and winning seemed a lot more important in that moment than looking like a tough guy.

Just as I predicted, all four Japanese men threw *rock*, making me the new champion. The crowd went wild and so did I! I held those longhorns over my head in victory, and the applause was deafening.

It was one of the best nights of my life, but I can't describe exactly why. I was out of my element—the center of attention in a room of screaming Japanese people—but somehow completely at ease. I'd wanted to win, the crowd had wanted me to win, and I hadn't disappointed them. Looking back, I see it as emblematic of my entire experience in Japan. I was surrounded by people who believed I had a strategy—that I was wise—and they were rooting for me to succeed.

I thought a lot about that night, and it renewed my self-confidence. It may seem a little silly, since the rock-paper-scissors game is largely a game of luck, but that doesn't mean I didn't have a strategy. It wasn't a stretch to see my performance in the bar as an example of situational decision-making—exactly what I was being employed to do at the site. I didn't save any lives that night, but I did acquire an impressive set of longhorns, which I proudly took home at the end of that long, amusing night. Even today, my Facebook profile is a picture of me celebrating the win with those longhorns over my head.

The third time I went to Little Texas with my friends was more heartwarming, if not as exhilarating. Matt had asked Rick Perry, then the governor of Texas, to make the owners of Little Texas honorary citizens of the state, and this was the night they were to receive "citizenship."

It was a very special ceremony; Matt made several proclamations in Japanese about the unity and affinity of our two countries. Little Texas represented the common bonds that we shared, he said. In that bar, we were one people. When Matt finished, there wasn't a dry eye in the house. It was as if he had expressed in words what all of us had been feeling throughout our experience together.

Each time I heard that Japanese band sing "America," I saw the emotional impact it had on everyone in the bar. You could tell that the Japanese people understood America, loved it, and felt as though we were connected in a special way. After my nights at Little Texas, I had a pretty good understanding of what the Japanese people were thinking and feeling, including their skepticism toward their own government and TEPCO.

Based on what I learned, I wrote a speech to present at the next Kantei meeting. I wanted to share my experiences and help the Japanese government officials understand a bit more about the attitudes of their own people. I ended my speech with a simple statement: "The Japanese people are counting on our joint leadership to resolve this issue. Together, I believe that we can."

It seemed that few of the men in that room had experienced "normal" life as I had on my visits to Little Texas. (To be fair, most of them had been spending every waking hour working to resolve the crisis and had had little time for such things.) I wanted to do more than just share what I'd seen in the "real world." I wanted to encourage the Japanese officials to get out of their offices and meeting rooms and into the communities themselves.

Solving a problem requires multiple perspectives. Hosono understood that, but he was exceptional. Understanding the needs and desires of real people can help to renew your passion for problem-solving and put a human face on large logistical issues.

In my speech, I mentioned how one can get an almost spiritual boost by speaking to those who truly need help. Enthusiasm and drive to resolve a problem are renewed, just as they were for me at Little Texas.

My visits to that basement bar changed my view of the Japanese people as fundamentally like us, with a love for their own country and for ours, and an appreciation for all we were doing to help them. Their deep-seated doubt about their own leaders threatened to spread like a contagion (as it had done in Hungary during the Paks crisis) and needed to be assuaged.

We saw a similar situation in the wake of Chernobyl. I believe that lack of faith in the government contributed to the fall of the Soviet Union: The people lost all confidence that their government could protect them. When that happens, the government's authority and credibility disappear. We were headed toward that same situation in Japan after Fukushima, if the government didn't pro-actively consider the needs of its people.

And don't think that such a crisis of confidence could never arise in the States. After Hurricane Katrina, many lost confidence in their local and national governments' ability to help and protect them. Losing a degree of confidence may not always take down a government, but it can certainly destroy the legacy of an administration. In Japan after Fukushima, Prime Minister Kan was the face of the government. He willingly took the fall for the accident and the poor response to it, whether this was justified or not.

As my experiences at Little Texas reminded me, measuring the pulse of the people can reinvigorate your work ethic and passion. Creating a strategy is impossible unless you take the time to under-stand the players. I came to believe that if government officials took the time to know more about the people they are sworn to serve, their approach would be better tailored to serve them.

When I walked out of Little Texas, I was more driven to work for the Japanese people—to protect them and honor them. I had laughed with them, hugged them, celebrated, and drunk into the wee hours of the morning with them. All of those seemingly insignificant things are important in Japan. They tied us together in a beautiful way.

From the beginning, I'd been committed to helping the Japanese recover, but my experiences with the people of the affected area deepened that commitment. They became a part of my life, and I began to love them just as I do my native countrymen. That affinity drove everything I did from that point on. I felt linked to that nation; they were, in a sense, *my* people, and that made it easier to help them.

CHAPTER 21

THE BIG TAKEAWAY

We cannot ever let this happen in the United States.

From my first trip to Fukushima through all my subsequent ones, the most persistent thought I had was, *we cannot ever let this happen in the United States.* We can't permit a region anywhere in our country to become contaminated and then attempt to evacuate it. The fear and anger it would engender would be immeasurable, and the backlash, I shudder to think about. Every aspect of such an event would be god awful. Even a seemingly small component, such as reentry for the purpose of collecting prized possessions, would be a logistical quagmire. I am not at all certain that we've given sufficient thought to the organization of this process, or any other related to a mass evacuation. Yet, somehow, the Japanese managed it.

Over the course of months and multiple trips into that evacuated zone, I could see that residents had gone in and tried to preserve their homes in small, improvised ways. In the aftermath of any catastrophe, a substantial percentage of people want to return home to live. (On the flipside, some 60 percent would never consider coming back.) It's difficult to return. Homes left abandoned and exposed to the elements for long periods quickly become uninhabitable. I noted that the blue-tarp roofs that I'd seen outside of the evacuation zone were absent inside the zone. Clearly, no one inside the evacuation zone had had time to take preservation measures.

The rice paddies I saw were unsalvageable and sprouting weeds. Knowing that many of the people in this area were subsistence farmers, I wondered what would become of them if they were not able to grow rice in the future. Over and over, I was forced to question whether our nuclear industry people in the United States fully understood what had happened in Japan, how far-reaching, how all-encompassing the consequences were. *Could anyone get an accurate impression from thousands of miles away?*

Fortunately, in 2013, all the American chief nuclear officers (CNOs) traveled to Japan as a personal mission to learn about the accident first-hand. Many of the CNOs felt that the trip was life-altering for them. Of all the countermeasures that the U.S. nuclear fleet instituted after the Daiichi and Daini accidents, the CNO trip was probably the most important.

At several points during my assignment there, I became somewhat frustrated at my own government, which kept pushing the stretched and depleted Japanese to act faster and take on bigger tasks. At one point, I chastised my superiors to *give the Japanese a break; they have much more to deal with than just the reactors.*

As I witnessed on successive site visits, conditions throughout Fukushima Prefecture progressively got better. By the end of 2011, much of the road damage was fixed. The cleanup at Daiichi, which had progressed slowly at first, picked up speed as the workers learned how to address radiation concerns.

From July through December 2011, the Kantei meetings continued, though we held them less frequently as time passed. We faced many one-of-a-kind issues that had never been faced by leaders in one place, at one time. There were agricultural, environmental, and social concerns, and the list went on.

In Japan, they refer to 3/11 as the "Triple Disaster" (earthquake, tsunami, and nuclear accident), but I consider it the "Quintuple Disaster" because, in reality, five crises were unfolding simultaneously: the earthquake and the tsunami, plus nuclear, social, and policy crises. The social and policy problems were part of a "system failure" surrounding the accident, similar to the one we experienced during our own Hurricane Katrina—i.e., local, state, and federal policy all failed.

During the half-year period following 3/11, the Japanese faced unimaginable and seemingly unending challenges. We worked with them to find solutions and help any way we could. It was a slow process to get the reactor cores cooled to the point of cold shutdown, but they achieved it officially on December 21, 2011.

Much of our work involved mitigating the damage to the evacuated areas, especially to crops and livestock. We had to reverse the voluntary early departure for dependents, adjust and eliminate the evacuation advisory, help complete the investigation into the cause of the accident, and develop diagnoses regarding the condition of the reactor cores.

In addition to all that, the Japanese asked us to provide advice on their initiative—called "Japan Is Open for Business"—to restore normalcy in the country. Our embassy was certainly interested in restoring Japan's businesses and safeguarding our relationships with them, so we helped and advised as best we could.

At the site itself, technical issues we worked on included the health of the workforce; long-term measures for removing radioactive fuel and the introduction of remote controls for removal of radioactive material; establishment of a permanent cooling system; accumulated water treatment; environmental impact assessments; and, importantly, stoppage of groundwater intrusion into the plants.

The Japanese collected radioactive debris from the plant, including all the trees that had been contaminated. They removed spent fuel from Unit 4 and, as of this writing, were still working to remove it from Unit 3. It will be several more years before they can complete the removal of spent fuel from Units 1 and 2.

Access to the reactor buildings is still either impossible or difficult. The Japanese have made great advances in their efforts to devise and employ robotics to enter the buildings, so that they can analyze and characterize the damage there. It will be more than a decade before they can start thinking about ways to remove the molten fuel from inside the reactors, which they had begun to find in December 2017.

Figure 12. The author and Goshi Hosono, special assistant to the prime minister

Over the last six months of 2011, we gradually reduced the frequency of our interactions with the Japanese and slowly reduced our staff. In the end, two of us were the main points of contact.

On the day that the Japanese declared all the reactors in cold shutdown, we held our last Kantei meeting—a ceremonial one. The Japanese, who value greetings and goodbyes quite highly, graciously hosted a very special thank-you party. I will forever appreciate my time spent with them, and their friendship.

Now that we have looked at the accident in detail, as well as its immediate and long-term aftermath, it is time to think about what caused it in the first place (and what causes various accidents of this kind). The first steps on the path toward the accident were taken before the reactors were even built. At some point, decisions were made about what kind of reactors should be constructed and where they should be placed. Those early decisions set a course of *path dependence* that ultimately led to the destruction of those plants.

In pre-war Japan, there were approximately 150 companies of various kinds competing to provide electricity throughout the nation. After the war, for the sake of efficiency, the Americans decided to reduce the number of electrical companies to nine or ten, and prevent them from competing. (In general, I believe competition to be a good thing, fostering excellence.) When the Japanese decided to add nuclear power to its utility system, the ancient imperial military sites seemed like good places to put the plants because, for the most part, they were in remote industrial areas and had switchyards (large electrical substations to support the electrical production facility).

At Fukushima Daiichi, before the plant could be built many feet of soil had to be removed to get down to bedrock. For that reason, the footprint of the site was very small, and the design had to be compact. Excavation made it vulnerable to tsunami. Over some objections, the architects chose to put the emergency diesel generators down in the basement of the facility, which made them vulnerable to flooding.

The next influence that contributed to Daiichi's path was the Chernobyl accident. During the Chernobyl accident about 6,000 children developed thyroid cancer from ingesting radioactive iodine. When the Japanese studied the Chernobyl accident, they grew determined not to put their own children at risk of cancer in case of an accident. A noble goal. Accordingly, they altered operational procedures. In case of an accident, the venting of Primary Containment would be delayed. Essentially, they'd "bottle up" the containment as long as possible, i.e., late-venting, giving operators time to establish water injection and core cooling, thus limiting the release of radioactivity from the facility. They chose against what is known as "early venting." As I discussed earlier, releasing a small amount of radiation early in an accident is better than a large release should the Primary Containment fail. During the Fukushima Daiichi accident, that late-venting procedure was followed—and the site superintendent was also given authority to evacuate an area of 1.5-km around the plant. Ultimately, the procedure and physical damage led to the explosion of Unit 1.

When the prime minister heard that the operators were conducting an evacuation, he directed the government to expand the evacuation area to nearly two miles. According to the "law of circles," this widened zone took at least twice as long to evacuate. The process of ensuring that the town was evacuated delayed the venting of the containment, and this, *theoretically,* caused a cascade effect. During that delay, the Unit 1 reactor building exploded from the hydrogen buildup. That, of course, affected the progress they had made on the other units because the explosion damaged equipment they needed and the release of radioactivity imperiled the workers and heightened their fears.

After an accident, everything down to the smallest detail is discussed, kicked around, brainstormed, and picked apart. Much of what I've just laid out in this discussion could be considered "counterfactual" or theoretical, but it's vital in my industry—every industry, really—to constantly study our mistakes and try to learn from them. A strong case could be made that the venting delay was extremely counterproductive and probably contributed to the severity of the accident. Decisions made years before contributed to the inevitability of the accident.

The design, the location, the operating procedures, all these factors lay in wait, so to speak, for the events of March 11, 2011. The challenge is this: We can't know whether we've made the right decision until it is tested in some way. A lack of testing can lead to a false perception of safety—until the next accident. Only then can we look back and see the mistakes we may have made.

How does this affect us going forward? We have to examine the decisions we are making today, in response to the Fukushima Daiichi accident, and try to predict what effect they might have on the next accident—next week or thirty years from now. It is imperative that we get better at learning tomorrow's lessons today.

Every organization, nuclear and otherwise, must maintain a fundamental belief that an accident could happen, and plan for it accordingly. There must not be a "myth of safety." We respect the heroes of Browns Ferry, Three Mile Island, Fukushima Daiichi and Daini, and countless lesser accidents, but our goal must be to avoid the need for any more heroes.

The events of 3/11 and its aftermath knocked Japan to its knees. Understandably so. What the disaster didn't touch was the Japanese sense of pride. America and many other nations stepped in to help Japan until she could rise again. Slowly at first, then rapidly, the

Japanese recovered and rebuilt their world so that it was even stronger than it had been before the attack. Having faced the greatest forces of nature and physics imaginable, at least outside of full-scale war, the Japanese demonstrated their indomitability as individuals and as a people.

That said, I consider the social disaster caused by the events to be the fourth of the quintuple crises that befell Japan. In the face of a serious loss of confidence in the industry, Japan is still struggling to restart its nuclear power plants. Currently, only about seven among its 44 or so remaining reactors are operating, and some will never be restarted due to the results of environmental lawsuits.

The fifth element in Japan's quintuple disaster is the nuclear-safety policy crisis that Fukushima revealed. Before the accident, there were practical and policy imbalances in the industry. Utilities held a great deal of power over the regulator (NISA), and controlled the technology. Over the decades, a "myth of safety" took hold, and necessary safety improvements and policies were never implemented, or even discussed. There was no notion that an accident *could* happen, so there seemed to be no reason to change. If an engineer said the seawall needed to be higher, he was disregarded. Even as plants across the globe were being made safer, Japanese plants stagnated.

Before the accident, there was an improper balance of power in place. The authority of the regulator was weak and the responsibility for safety was vague, resting mostly on the vendors who maintained the reactors. Today, that balance has shifted. The regulator is powerful and has the ultimate authority and responsibility for safety. This sounds like a vast improvement, but I wonder if it is really the healthiest way of functioning. My own belief is that safety should rest primarily in the hands of a competent utility, with oversight provided by the regulator. The way it is now, the regulator simply demands that *all* risk be eliminated, regardless of practicality or cost. Without public debate and consensus to determine an *acceptable* level of risk, the regulator is reluctant to allow plants to restart at all. Until they resolve this new policy deadlock, it will be difficult to move forward on nuclear energy. The Diet must step in and resolve both the social and policy crises I've described, and provide sufficient oversight to the regulator.

The good news is that life is slowly returning to normal in Fukushima Prefecture. No one died from radiation exposure during the accident, and cancer rates are not expected to rise appreciably. Nuclear utilities in Japan and internationally have been able to take away many valuable lessons, and TEPCO has turned its focus to improving safety culture in whatever way it can. I call it the "new TEPCO," as it has changed from bottom to top.

There are other successes to note. Japan's version of INPO, called JANSI (Japanese Nuclear Safety Institute), is taking root, and is proving itself more independent and more competent than NISA. An American, William (Bill) Webster is the chair for this organization. Utilities have started independent nuclear safety advisory or review boards. Plants that are operating are doing so safely.

During the immediate aftermath of the events, TEPCO was understandably confused. Someone once described the situation as, *like trying to solve a murder without having access to the crime scene.* They did the best they could under the circumstances. Normal people engaged in genuine heroics during the response. Today, TEPCO has undergone a dramatic reform based on the lessons it learned. I trust their new safety culture and think that they will be a premier operator of nuclear plants moving forward.

Over the years since the accident, I've reflected on the extreme-crisis leadership issues that arose, and have written about them throughout this book. Here are some final thoughts on how these incredible leaders led their teams through an amazing crisis.

How can you, as a leader, make sense of an extreme crisis? How can you get your mind around such a complicated situation, so that you can make decisions? We are only human. We cannot foresee all the possible moves on a chess board, and we can't predict all the possible outcomes in a crisis. *Sensemaking* is about taking cues from the environment and assembling them into a picture of what is transpiring. This happens when you create an "enacting" organization structure; that is, you lead your way through an event by taking actions to help understand it. During an extreme crisis, your cognitive abilities may not be up to the task of making sense of the events unfolding. As a result, you may find yourself oversimplifying the events to keep from experiencing overload or panic. Your actions can generate either a better or a worse understanding of the events and can change their progression either positively or negatively. The best outcome is that your actions both improve your understanding of the events and change the outcome positively.

A major challenge for leaders during a crisis is to develop a shared understanding of the situation that makes sense to most people, given what they know at the time. You must remember your role as a leader and remain neither too pessimistic nor too optimistic. It may be tempting to issue overly optimistic statements or explanations simply to justify your own decisions, but this can lead to serious blind spots and overconfidence in a picture of your own creation. Wisdom, based on constant analysis of input and a healthy sense of doubt about everything you hear, can help lead to shared understanding of the situation. My own healthy sense of doubt came into play often at Fukushima, as you have read. By constantly updating your mental model and holding on to your skepticism, you can adapt and innovate as needed, and lead others to do the same.

The Fukushima Daiichi and Daini leaders I have written about embodied the principles of sensemaking. Both Izawa and Masuda acted decisively until events overtook their sensemaking abilities, then consciously took a step back to reassess the situation and regain their sense of it. When unexpected challenges came hard and fast, they responded to a shifting reality problem by problem, then *acted* their way toward sense, purpose, and resolution. They never lost focus, calmness, or hope.

An unexpected crisis disrupts the familiar. Slowly we regain familiarity by taking action and then stepping back for reflection. With each new problem they faced, these leaders recalibrated, iteratively creating continuity and restoring order. They acted their way into a better understanding, and as they did so, some things became more certain. Other things became less certain. They went through a lot of trial and error and avoided overcommitting to one exclusive success path.

There is a fine line between rushing *into* to a solution and working incrementally *toward* one. Masuda demonstrated this when he calmly presented his people with the uncertainty of their situation, allowing them to embrace the changing nature of it and the need for adaptive solutions. So, even when he had to shift strategies, they understood that, for good reason, he was leading them down an alternate path to success.

I viewed the leaders in place during the Fukushima accident as victims as well as leaders, acknowledging that their experiences had tangible psychological effects on them. This inspired my sympathy toward them in the moment and in speaking with them afterwards. Leaders who come in to lead after an event approach matters differently from those who experienced the event personally. It makes sense that they lead differently. As outsiders coming in, we couldn't help but have a different view from those who'd found themselves in the throes of the disaster. I understood that it was important to show deference to them for this reason.

Bridging communications challenges is what I call *teamsense*. Teamsense is sharing the situation with your team so that everyone can engage in their own sensemaking. Thus, many teams of workers, undertaking a variety of tasks and challenges, can arrive at a common understanding, and can adapt to each problem. They can repeatedly act with decisiveness through each successive crisis.

Teamsense and sensemaking depend on effective communication. For example, the use of whiteboards to share information at Daini reduced uncertainty and fear and gave substance and meaning that the workers felt to their cores. I found it humorous when Masuda told me that he'd needed to make the information convincing even when it wasn't at all. Whiteboards turned out to be a very good way to communicate the inevitable adjustments that come from working through monumental challenges such as these. Team understanding promoted team buy-in, and uncertainty was replaced by meaning.

At Daiichi, it was extremely difficult for the ERC and the field to communicate. This threatened worker safety and interfered with the vital flow of accurate information. To offset the problem, Izawa gave detailed directions and required workers to repeat back his commands. This ensured that when they set off on a task, they understood his expectations perfectly.

Sensemaking and good communication foster good decision-making, but there will always be setbacks in a crisis. Managing setbacks takes skill and, above all, flexibility. Because Masuda presented the realities of the situation to his people calmly, the workers embraced the unpredictable nature of their tasks. Izawa, Inagaki, and Masuda all faced setbacks, but through effective leadership, they *enacted* their way through the problems.

As leaders develop a shared mental model (*sensemake*, if you will), they can make decisions. Of course, it is best to make decisions based on facts but, because of the nonlinearity of extreme events, the likelihood of having all of the accurate facts is minimal. In the immediate aftermath of an extreme event, decision-making is naturally fraught with difficulties and less than optimal.

You might think that with more information comes more clarity, but sometimes, with more information comes more confusion. In any case, the best remedy available to leaders when few facts are available is to enact basic rules (usually safety rules) and provide fundamental instructions, then empower workers to confront and process uncertainties in the field as they find them. Leaders who allow their teams to do their own sensemaking—as the best leaders at Fukushima did—make it possible for those on the ground to cope with the emerging reality around them, assessing risk as they go.

Never forget the importance of vetting the facts and assessing their source. This must be done while keeping pace with the crisis, which may be unfolding with great speed. In the end, the leader must make decisions with or without an optimal number of reliable facts. Often, leaders reach out for advice from technical experts—but this can slow the speed and affect the efficacy of the decision-making processes. Experts may not have access to the technology they need to assess matters, or may simply be wrong about the situation at hand. At Daiichi, experts often proved ineffective as they were not in the best position to counsel those in the trenches.

The workers at Fukushima experienced the limits of human emotion and physical capacity. Death-related anxiety—when the prospect of death becomes real—can alter a person's ability to think clearly and effectively. When thoughts of death are subliminal, people tend to generate positive thoughts to overcome their fears of mortality. *It's not that bad,* they think. *It will be OK.* But when the possibility of death looms as a reality, the mind narrows and cognition becomes limited. Morals change as well. Acts of defiance are common. The personal flight instinct and the shift from logic to intuition occur. Understanding all of this, the behavior of the leaders and workers at Fukushima seems even more amazing and admirable. Izawa, Inagaki, and Masuda somehow overcame these instinctive reactions in themselves, and provided the kind of leadership that allowed workers to overcome them as well.

Emotions play a significant role in extreme-crisis management. In fact, it's the emotional piece that makes a crisis "extreme." Thus, the character of a leader may be the most important element of leadership in a crisis. When I refer to *character* in a leader, I do not mean personality or values. Character in this context involves the leader's mastery and hands-on experience, and how she/he can contextualize these across the organization. I enjoy this quote by the southern novelist James Lane Allen: "Adversity does not create character, it reveals it." Certainly, the Fukushima crisis revealed the character of many leaders. Although extreme leaders are highly trained, skilled, and knowledgeable people, they are human, not robots. They are subjective, emotion-laden, and susceptible to joy and sadness, even as they attempt to lead wisely and well. This is especially true in a commercial (nonmilitary or first responder) environment.

Extreme-event leaders must understand that all eyes are on them in the immediate aftermath of the disaster, and remain calm. They must understand that their staff's trust has reset to almost zero and must be rebuilt with every decision, emotion, and action they demonstrate. After the explosion of the Unit 3 reactor caused injuries, some workers returned to the ERC and yelled at Yoshida, "You liar, you liar." Their level of trust in Yoshida had reset to zero, and it was up to him to rebuild it. Leaders must be aware of the potential for defiance when planning and training under extreme conditions. In an extreme crisis, defiance can be a natural reaction to fear. We don't plan for defiance in our emergency training.

It's impossible to hear the stories of Izawa, Yoshida, Masuda, Inagaki, and others without feeling the gamut of emotions they experienced. Soon after the earthquake, and specifically during the tsunami, leaders and followers alike crossed a threshold of emotion. A normal reactor shutdown takes place in a controlled environment; the operators are well trained to follow procedures they are familiar with and do so with trust and bonding. Safety culture changes radically during an accident. When the tsunami struck, power was lost. There were no lights, no controls, and no instrumentation. The operators could not determine radiation levels, and they crossed a threshold into uncertainty and fear.

Throughout the Daiichi event, operators experienced the gamut of emotions. They felt honor, shame, guilt, embarrassment, anxiety, pride, hopefulness, and even humor. These emotions conditioned their responses and impacted the outcome of the event. At Fukushima Daiichi, as death anxiety caused cognitive narrowing, the operators became more defiant and resorted to their intuition for self-preservation. In many cases, this caused them to defy their leaders. The young operators in the control room felt vulnerable and hopeless and wanted to evacuate. Only after Izawa assured them that he would protect them against "outside interference," begged them to stay, and apologized to them, did they agree to remain. In some cases, Izawa tried to offset constructive defiance by acknowledging the hazards and appealing to their sense of duty. In another case, he withheld disturbing information, fearful that his workers would panic.

The quintuple disaster of 3/11 was a "transboundary" crisis. That is, the accident reached around the world; it was not limited to one Prefecture or even the nation of Japan. In our globalized world, it was the first live, web-streamed nuclear accident and it carried implications that included challenging the strength of the U.S.-Japan alliance. Foreign governments closed their embassies. Had the U.S. Embassy closed, it would have had a major effect on our strategic relationship with Japan. This "transboundary effect" is an example of the "coupling" of modern society. Natural events caused the Fukushima Daiichi nuclear plant accident; it was then complicated by humans and cascaded into a global social crisis. Leaders must control panic and mitigate the impact of a crisis at the earliest point in its progression to keep a "routine" crisis from turning into an extreme one. Transboundary effects can exert tremendous pressure on political leaders to make social decisions that could be at odds with technical realities.

Crisis leaders tend to just consider the technological issues involved in "solving" an event without looking at other factors. They then manage the consequences. Meanwhile, the media is disseminating the technological situation through a social prism, often inflating it by posing "what if's" and postulations by pseudo-experts. As the amplification increases, more sociological pressure is placed on the politicians, much of it counterproductive to mitigating the crisis. At Fukushima, the general fear that highly radioactive water would reach the ocean led the Japanese prime minister to insist that this be prevented at all costs. The result was a dangerous delay in the safe shutdown of the reactors, and the ensuing explosions and radioactive plumes.

As important as what these leaders did is what they did *not* do during the crisis. They did not submit to panic and they did not overcommit to one plan. They shared their plans freely and did not "overlead." They acknowledged the evolving reality openly and shared the burden of uncertainty and doubt with their workers.

There were some questionable leadership missteps, no doubt. Yoshida later admitted that he'd missed a signal from the operator about the isolation condenser operations. Many people have argued with Masuda's decision to ration the bento boxes. Izawa's decision not to share the conditions of the reactors with his operators might be questioned. On balance, however, I would argue that the leadership's decisions, directives, and behaviors were first rate and more than admirable under the circumstances.

This accident was truly a system failure, stemming from the various causes I have already outlined. The operators' comparison of themselves to American veterans of the Vietnam War is an apt one. They'd taken heroic action, yet their country viewed them as villains. That is one of the reasons I wanted to tell this story. These people are heroes, nothing less. It was the system that failed.

My hope is that this book brings to life the incredible events of a dire time, and illustrates some of the crisis-leadership principles that came into play. As a leader, you may never face the kind of challenges that the heroes in this book faced; I hope you don't. But the takeaways from their story may very well be applicable to your own life and work. Furthermore, I can't help thinking that making sense of the efforts of those who led Japan through the crisis honors their efforts in a new way.

CHAPTER 22

JUNE 2014 TRIP TO FUKUSHIMA PREFECTURE

*I asked him, "Why didn't you use the other lane?" and he
looked at me dumbfounded, as though the idea had truly
never crossed his mind.*

Three years after the accident, having officially retired from the NRC, I went back to Fukushima Daiichi and Daini and toured the prefecture. By that point, I had been to both reactor sites many times and had seen progress on each occasion. Sadly, Yoshida died of a preexisting ailment about a year after the accident. Masuda, the leader from Daini, took over the recovery of Daiichi.

Traveling around Daiichi three years after the accident, I could see the hand of Masuda everywhere. Having organized an admirable crisis response, this skilled leader was now effectively managing the colossal radioactive cleanup.

I had never previously had the opportunity to spend much time talking with locals in the Fukushima Prefecture, so I looked forward to doing that on this trip. My plan was to meet with business professionals, mayors, officials, and regular people to discuss how things had changed since my last trip to Daiichi, a year prior. I traveled with an American professor friend of mine, Kyle Cleveland and we hired a Japanese freelance reporter to serve as our guide and interpreter. She picked us up at the Fukushima train station and happily drove us around the area.

This trip offered a very different experience from previous ones.

We drove through a desolate evacuation area that had remained empty, though its boundaries had shrunk over time. The earthquake had severely damaged Highway 6, but it had reopened, once again providing a direct link between Iwaki Village and the Sendai area. Traveling along it, through the evacuated zone, we saw some rice paddies where the soil had been removed and cleared of cesium. The contaminated dirt was stored in bags that were stacked around the edges of the paddies. Every aspect of life for those in and near the contaminated area had been affected by the 2011 disaster.

Some farms were operating again, but were growing rice primarily for personal consumption. No rice from that area was being sold commercially, due to contamination concerns. In fact, in some food stores in Japan, harvest maps sprouted up in front of vegetable stands showing where the crops were grown, thus showing that they were not grown in Fukushima Prefecture. Clearly, the locals didn't have the luxury of worrying about this—they needed to eat. A few other businesses were open as well, but these were rare and life was anything but normal. When surveyed, only 40 percent of the people from the affected area had said they intended to return, and most of those were older people, sentimental about their homes and ill-equipped to start new lives. Younger people knew there were no jobs for them in Fukushima Prefecture and moved on. We did note some signs of recovery. At one point, we got stuck in a traffic jam and nearly missed our train. I fretted about it until I realized what a positive sign it was.

One of our most memorable experiences in Fukushima Prefecture was our trip to the small town of Namie, northwest of Daiichi. This was the town that had been at risk during the venting of Unit 1, and was the last to be evacuated, delaying the venting of Unit 1. We had the opportunity to meet the former mayor of Namie, Tamotsu Baba and had a great conversation. Understandably, he is no longer a supporter of nuclear power usage. He explained that they didn't understand the radiation monitoring system that was in place at the time (SPEEDI). It printed out page after page of complicated measurements, which were confusing and incomprehensible to them. As he showed us photos of the evacuation process, I noticed that on two-lane roads, only the outbound lane

Figure 13. The author and the mayor of Namie Town, Tamotsu Baba

had been used. Since everyone was going in the same direction, wouldn't it have made sense to use both lanes? The mayor looked dumbfounded at the suggestion, as though the idea had never crossed his mind. A simple shift in perspective could have significantly improved the entire process.

Other missteps could be seen clearly in retrospect. Around a dozen critical-care patients had died during hospital evacuations. Some of them might have been saved if elevators had been used, but hospital workers had followed their training and avoided them after the earthquake because they hadn't yet been inspected. Obviously, in this case, no inspector was going to arrive, and using them would undoubtedly have been worth the risk.

After Namie, we spent some time in Fukushima City, where we spoke with a pediatrician about the current health of residents. Before the accident, the Fukushima Prefecture had had the fourth-lowest incidence of cancer in the country. Despite the increased risk of it since the accident, the statistics were good, thanks to their dedication to prevention and treatment. The doctor told us that they were receiving extra funding for wellness checks of children and families in the region. This boosted the overall health of the population, he pointed out, because during the screenings, many other health conditions were identified. We did find an unusually elevated rate of childhood obesity in Fukushima Prefecture, probably because children were being discouraged from playing outdoors for fear of contamination.

We spoke with elected officials from other major cities in the region, such as Koriyama, but those discussions were not particularly fruitful. The leaders' anger and disdain for the nuclear plants and the damage they had caused clearly clouded their responses to our questions.

In some of the larger cities we visited, the differences in evacuation zones between the Japanese and the Americans arose in the discussions. Apparently, some of the Americans' boundary lines had split neighborhoods in half, dividing streets into "dirty" and "clean" sides. The inconsistencies in the process made people fearful of the evacuees themselves, which made sheltering people a problem. During that chaotic time, fear and emotion ran high and people questioned one another's motives. It all made the process very challenging.

Throughout Fukushima Prefecture, radiation meters were installed to inform the public of the local radiation level and provide reassurance that the area was safe. But clearly, the accident had changed people's perspective forever. There was an instance where I was wearing a radiation-measuring device in a local community. An elderly woman asked what it was for. When she heard the word *radiation,* rather than listening to my simple explanation, she practically vaulted in the other direction due to her fear.

A forty-year-old entrepreneur we met had owned a trash hauling business, among other ventures, before the disaster. The evacuation and slow return devastated his companies. Without people in the area, there was simply no trash to haul, and his trucks and supplies were sitting idle. Rather than giving up, the man applied for and won a government contract to conduct decontamination work in Fukushima Prefecture. Intrigued by the work he and his crew were doing, we visited them out in the field, where they were operating front-end loaders, weed eaters, shovels, and large trucks to clear out rice paddies. They would shovel out a span of six feet from the side of the highway, then another six feet up the hillside, scraping off all the potentially contaminated dirt and hauling it to a centralized storage facility.

Figure 14. Radiation meter in a city park

After observing this, we visited the disposal area. It was an amazingly poignant sight—acres and acres of dirt piles all covered with green tarps that somehow made it look like a lush green field. It was anything but. The entire area was fenced in and being monitored for radiation levels.

The entrepreneur of this dirt-hauling business was a friendly character who hired a lot of young men for his crew, given that it was a rather physically demanding job. We talked and joked with some of the boys, and I even arm-wrestled one of them. When I

Figure 15. Contaminated soil disposal site

went out to the parking lot, I saw that most of them owned fancy SUVs with all the bells and whistles. Clearly, they were being paid well for their work and living good lives in the post-Daiichi period. I was most impressed with the owner, who appeared to be a great mentor for the boys, teaching them much more than just how to haul dirt. He was setting a standard for hard work and dedication that I believe was having a positive, long-term effect on them.

I wanted to know how they had chosen the six-foot standard of clearance, but it was difficult to get a basic answer; there didn't appear to be much of a scientific basis for it. The crew told me that six feet was as far as they could reach while standing in the road. (Going into the woods would have increased their exposure to the contamination.) They understood that it would be impossible to clean the entire forest, so their goal was to minimize the background radiation levels on the roads and directly adjacent land, because that was where the most people would be affected. This may not have been entirely grounded in science—in a way, it was another "optical solution"—but it certainly served as a means of showing progress. It also provided work at a time when people needed it. In addition to hauling dirt, these boys were delivering hope to their community.

Following the tsunami, earthquake, and Daiichi disaster, many people wanted to travel up to Fukushima Prefecture and volunteer in the relief efforts. A few months into the relief process, however, the locals began rejecting help from these volunteers because they were essentially taking jobs away from locals who needed them. The government was paying locals to contribute to the cleanup, but volunteers were doing the same thing for free. While the sentiment and intention of the volunteers were appreciated, their actual presence was not.

A few miles from Daiichi is Tomioka Village, considered "ground zero" for the disaster. Before 2011, it had been a quaint seaside town with several resorts catering to visitors on vacation. The earthquake and tsunami destroyed it, then the nuclear accident covered it with a blanket of radioactive cesium.

It was tragic but fascinating to walk through the remains of that little village. Unlike other areas in Fukushima, where visible progress had been made in the cleanup, Tomioka remained as it was on that fateful day. Clocks visible in the rubble were stopped at the precise time of the earthquake. When and if it would be cleaned up was a mystery to me. Displaced locals had been permitted to visit—in protective gear—but not to stay. Our own visit there was possible only with the support of TEPCO, whose vans we were riding in.

Toward the end of our visit, I noticed a shrine near the waterfront where several people had gathered. I expressed my desire to meet them and express my sympathy for all that they had lost. As I set out to do so, one of the TEPCO officials shouted, "Dr. Casto! Dr. Casto!" I turned to see what he wanted, and what he said surprised me. "TEPCO officials can't go near the residents," he explained. "They blame us for all of this." Reluctantly, I returned to the van.

That trip provided me with a broader perspective on the disaster than I could possibly have had three years earlier, when the devastation was fresh and painful and the dangers still unquantifiable. This trip reminded me of the determination and pride of the Japanese people. We spoke to many from the prefecture, ranging from day laborers to elected officials, and our overall impression was that life was returning to normal. There were still extreme challenges to face, including the long-term environmental impact on the forests and soil, but work was being done to address these, move on, and grow. The resilience of the people and landscape of Fukushima Prefecture is inspiring, and the people there stand as a noble example of community, unity, and fortitude in the face of disaster. As we in America might put it: #fukushimastrong!

GLOSSARY

AOV
Air-operated valve

BWR
Boiling Water Reactor

CNO
Chief nuclear officer

CRD
Control rod drive

Daiichi (daa-ii-chi)
Fukushima Plant 1, Daiichi: *Dai* means "number," and *ichi* means "one" in Japanese ("number one"). Reactors 1–3 operating before earthquake, 4–6 shut down for maintenance before earthquake.

Daini (dai-nee)
Fukushima Plant 2, Daini, *Dai* means "number," *ni* means "two" in Japanese ("number two"). All four reactors were operating prior to the earthquake. Daini was at a higher elevation, which helped to keep some power supplies available.

DDFP
Diesel-driven fire pump

DOD
Department of Defense

DOE
Department of Energy

ERC
Emergency Response Center (located at Daiichi—0.5 mile from control room)

Fukushima 50
Operators who remained behind after Daiichi site evacuation

GDP
Gross domestic product

HPCI
High-pressure coolant injection

IAEA
International Atomic Energy Agency

I&C
Instruments and controls

IC
Isolation condenser (a passive system at Daiichi unit only)

ICS
Incident command structure

INPO
Institute of Nuclear Power Operations

JANSI
Japanese Nuclear Safety Institute

JET
Japan English Teaching Program

KI
Potassium iodide pills to protect thyroid

METI
Ministry of Trade and Industry

MOD
Ministry of Defense

MOV
Motor-operated valve

NISA
Nuclear and Industrial Safety Agency—replaced by the Nuclear Regulatory Authority (NRA)

NRC
U.S. Nuclear Regulatory Commission

RCIC
Reactor core isolation cooling system—an active system at all reactors except for Daiichi

RIC
Regulatory Information Conference

SBO
Station blackout (loss of all AC power)

SDF
Japanese self-defense forces

SLC
Standby liquid control system

SPEEDI
Environmental Emergency Dose Information system

SRA
Senior reactor analyst

SRV
Safety-relief valve

TEPCO
Tokyo Electric Power Company

USAID
U.S. Agency for International Development

WANO
World Association of Nuclear Operators

NOTES

INTRODUCTION

1. Established in the aftermath of World War II, the U.S.-Japan alliance was first conceived as the Mutual Security Pact (1952) and later replaced by the Treaty of Mutual Cooperation and Security (1960). In this agreement, collectively nicknamed "spear and shield," Japan agreed to provide U.S. forces with basing rights on its territory in exchange for security against external threats.

2. The Jogan earthquake, known formally as the Sanriku earthquake, struck the northern part of Honshu on July 9, 869. With an estimated magnitude of 8.4, it caused widespread flooding of the Sendai plain.

CHAPTER 2

1. Sean T. Hannah, Mary Uhl-Bien, Bruce J. Avolio, and Fabrice L. Cavarretta, "A Framework for Examining Leadership in Extreme Contexts." *The Leadership Quarterly* 20 (2009): 898. Accessed May 15, 2018, at https://digitalcommons. unl.edu/cgi/viewcontent.cgi?article=1038&context=managementfacpub.

2. Charles A. Casto, (2014), "Crisis Management: A Qualitative Study of Extreme Event Leadership." *Dissertations, Theses, and Capstone Projects.* 626. https://digitalcommons.kennesaw.edu/etd/626.

3. R. Dynes, (1974), *Organized Behavior in Disaster,* Columbus: Disaster Research Center, Ohio State University; R. Dynes, E. L. Quarantelli, and G. A. Kreps, (1981), *A Perspective on Disaster Planning,* 3rd Edition (Newark, DE: University of Delaware Press).

4. "*Tendenko,* as a tsunami evacuation strategy, advocates moving quickly to safety in anticipation of an imminent tsunami threat by disregarding others, even one's own family members. The concept developed within a disaster subculture along the eastern coast of Japan as a grassroots response to large repeated tsunamigenic earthquakes in which entire families perished. High death tolls in tsunami events were attributed to the desire of families to reunite before evacuating, thus losing precious seconds and minutes during which they could have fled individually, saving their lives. Following the Great East Japan Earthquake of March 11, 2011, *tendenko* once again became a topic of interest in the news media and among academic researchers and government

officials." James D. Goltz, "Tsunami Tendenko: A Sociological Critique." *Natural Hazards Review* 18, no. 4 (November 2017). https://ascelibrary.org/doi/10.1061/%28ASCE%29NH.1527-6996.0000254.

5. In academia; the theory is called *sensemaking*—establishing a mental model, a model that fosters action. K. E. Weick, "Enacted Sensemaking in Crisis Situations." *Journal of Management Studies* 25, no. 4 (1988): 305.

CHAPTER 3

1. "U.S. Navy to Provide 500,000 Gallons of Fresh Water to Fukushima," U.S. Navy website. Accessed at http://www.navy.mil/submit/display.asp?story_id=59318.

2. The Emergency Response Center (ERC) is located on the reactor site, about a half mile from the reactors. It includes a seismic isolation building where the emergency response personnel led the response efforts. It was a two-story, reinforced-concrete bunker designed by TEPCO to act as a shelter and command center in major emergencies. It had two air filtration systems to keep out radiation and was built on dampers, or giant shock absorbers, that would allow it to survive even the biggest earthquakes.

3. A design basis "accident" is defined by the NRC (nrc.gov) as: "A postulated accident that a nuclear facility must be designed and built to withstand without loss to the systems, structures, and components necessary to ensure public health and safety."

CHAPTER 4

1. *Cold shutdown* means that the reactor pressure vessel, which is made of steel, is below 100 degrees Celsius and is considered safe.

2. Instruments and Controls (I&C) are the electronic processes to operate components or to collect data from the reactor systems. I&C people conduct testing and maintenance of these processes.

3. The high-pressure coolant injection system (HPCI) is a steam turbine pumping system that injects a high volume of water into the reactor when the reactor does not depressurize.

4. The reactor core isolation cooling system (RCIC) is a steam turbine powered high-pressure water injection system. It is designed to inject water into the core during conditions when the reactor is at high pressure before depressurization. It is a smaller, complementary system to the High-Pressure Coolant Injection System (HPCI).

CHAPTER 7

1. A bento box is a prepackaged lunch that is popular in Japan.

CHAPTER 9

1. TEPCO Headquarters is in Tokyo near the prime minister's residence.

2. Scott Pelley, "Nuclear plant operator remembers Fukushima." *CBS Evening News* (March 16, 2012). https://youtu.be/IaqdeRyK0pM.

CHAPTER 10

1. "The history of the failure of war can almost be summed up in two words: too late. Too late in comprehending the deadly purpose of a potential enemy. Too late in uniting all possible forces for resistance. Beware not the enemy from without but the enemy from within." http://www.tankmastergunner. com/quotes-macarthur.htm.

CHAPTER 11

1. The Naval Reactors Program is the U.S. military version of the NRC. However, there are two distinct cultures between them. Naval ships cannot tolerate any release of radiation. For the commercial nuclear industry, with the larger facilities, and commercial requirements, radiation exposures are managed.

2. The Institute of Nuclear Power Operations (INPO)—an independent, membership organization that promotes the highest levels of safety and reliability in the operation of commercial nuclear power plants. Https:/INPO.info.

3. The World Association of Nuclear Operators (WANO) unites every company and country in the world that has an operating commercial nuclear power plant to achieve the highest possible standards of nuclear safety. Https:/ WANO.info.

4. There were roughly 153,000 Americans living in Japan (plus tourists) at that time, of whom 103,000 were U.S. military personnel and Department of Defense dependents (family and military contractors).

5. After the Chernobyl accident in 1996, the Japanese government invested considerable resources in its implementation.

6. Refer to the June 17, 2011, letter from NRC Chairman Gregory B. Jaczko to Senator Jim Webb for information on assumptions used in recommending a 50-mile evacuation for U.S. citizens following the Fukushima Daiichi nuclear facility events.

CHAPTER 16

1. Yokosuka and Atsugi are about 45 miles from Tokyo, about 20 miles apart from each other, and about 200 miles from Fukushima Daiichi.

2. Kyle Cleveland, "Significant Breaking Worse," *Critical Asian Studies*, 46:3 (2014), 509–539, DOI: 10.1080/14672715.2014.935137.

CHAPTER 18

1. By October 2012, the Japanese Reconstruction Agency identified 1,000 disaster-related deaths that were not due to radiation-induced damage or the earthquake or the tsunami. http://www.reconstruction.go.jp/english/, based on data for areas evacuated for no other reason than the nuclear accident. About 90 percent of deaths were for persons above sixty-six years of age. Of these, about 70 percent occurred within the first three months of the evacuations. (A similar number of deaths occurred among evacuees from the tsunami- and earthquake-affected prefectures. These figures add to the 19,000 that died in the actual tsunami.) The premature deaths reported in 2012 were mainly related to the following: (1) somatic effects and spiritual fatigue brought on by having to reside in shelters; (2) Transfer trauma—the mental or physical burden of the forced move from their homes for fragile individuals; and (3) delays in obtaining needed medical support because of the enormous destruction caused by the earthquake and tsunami. However, the radiation levels in most of the evacuated areas were not greater than the natural radiation levels in high background areas elsewhere in the world where no adverse health effect is evident, so maintaining the evacuation beyond a precautionary few days was evidently the main disaster about human fatalities. http://www.reconstruction.go.jp/topics/20121102_sinsaikanrensi.pdf. http://www.reconstruction.go.jp/topics/240821_higashinihondaishinsainiokerushinsaikanrenshinikansuruhoukoku.pdf. http://www.world-nuclear.org/information-library/safety-and-security/safety-of-plants/fukushima-accident.aspx.

CHAPTER 19

1. A cooler title, I can't imagine. I jokingly complained to the NRC team that I didn't have a cool title. Before I knew it, our team created an equivalent title for me—USNRC Forces Japan Supreme Allied Commander.

2. I read an article that estimates the tsunami's force at just under one atomic bomb. Kenneth Chang, "The Destructive Power of Water," *New York Times* (March 12, 2011). http://www.nytimes.com/2011/03/13/weekinreview/13water.html.

3. The Hanford site is a decommissioned nuclear production complex operated by the United States government near the Columbia River in the State of Washington.

4. In 2014, I coauthored an article that details the leadership success of Team Masuda. See Ranjay Gulati, Charles Casto, and Charlotte Krontiris, "How the Other Fukushima Plant Survived," *Harvard Business Review* (July–August 2014). https://hbr.org/2014/07/how-the-other-fukushima-plant-survived.

INDEX

MEET CHUCK CASTO
IN PERSON

Do you need a compelling keynote or seminar speaker?

Dr. Casto delivers speeches and conducts training on disaster preparedness for organizations and on extreme-crisis leadership.

For more information, visit www.ChuckCasto.com.

• • •

"Chuck Casto's knowledge of extreme crisis leadership provides useful insights for any leader. Chuck's unique experience during the Fukushima nuclear disaster provides an up close and emotional picture of leadership during an extreme crisis. His message is an excellent, cogent, and dramatic look at what leadership is all about."

—LUIS REYES,
Retired, Executive Director for Operations at USNRC